A Compendium of Unique and Rare Spices

Kodoth Prabhakaran Nair

A Compendium of Unique and Rare Spices

Global Economic Potential

Springer

Kodoth Prabhakaran Nair
Villament G3
C/o Dr. Mavila Pankajakshy
Kozhikode, Kerala, India

ISBN 978-3-031-20248-3 ISBN 978-3-031-20249-0 (eBook)
https://doi.org/10.1007/978-3-031-20249-0

This Springer imprint is published by the registered company Springer Nature Switzerland AG
The registered company address is: Gewerbestrasse 11, 6330 Cham, Switzerland

I dedicate this book to Dr. Kannan Mavila, our son, who very diligently helped me gather all the relevant scientific material for this book, and also, helped me much, when I ran into computational difficulties. It has been a very great pleasure working with him on this important book.

Contents

Chapter 1
Saffron and Tulasi

Abstract *Saffron* The chapter begins with discussion of length, the origin, description and distribution of Saffron, worldwide. It goes further to discuss the various aspects of crop management of saffron, including soil requirements, crop protection, etc. It concludes with the discussion of the potential of Saffron, as a rare and unique spice of the world. *Tulasi* The chapter discusses Tulasi, the sacred Indian plant, worshiped by all Hindus. It highlights Tulasi's food value. Tulasi has also many pharmaceutical properties. These are discussed in the chapter.

Keywords for Saffron Origin · Description · Distribution · Economic viability
Keywords for Tulasi Sacred · Hindu · Mucilage · Food value

Saffron

Botanical Name: *Crocus sativus* Linn

Family: Iridaceae

Botanical name: *Crocus sativus* Linn

Description and Distribution

Saffron, the legendary crop of Kashmir, India, is an important food adjunct, that is, food flavorant and food colorant, and, is also reputed to possess several medicinal properties. The recorded accounts of saffron cultivation in Kashmir date back to AD 550, nearly four centuries earlier than its recorded cultivation in Spain by the Arabs around AD 961. The saffron fields, from mid-October to early November, when saffron plants are in full bloom, on a moonlit night, present a wonderful sight and a good source of fascination to many tourists. But surprisingly, its cultivation in Jammu and Kashmir in India does not seem to be based on scientific research and the growers have some superstitions, notions and methods of their own. Therefore, saffron-growing areas are those very localities where they have been since ancient times (550 AD) and are also their outdated methods of curing, drying and packing, which can be improved considerably.

Over 100 species of saffron genus *Crocus* are distributed worldwide, but only *Crocus sativus* has received some attention and is cultivated in several countries.

K. P. Nair, *A Compendium of Unique and Rare Spices*,
https://doi.org/10.1007/978-3-031-20249-0_1

Table 1.1 District-wise (Kashmir) area, production and yield of Saffron/ha

Districts	Area (ha)	Production (kg)	Yield/ha (kg)
Srinagar	218	1232	5.65
Badgam	215	903	4.20
Pulwana	2555	7793	3.05
Doda	34	205	6.03
Total	3022	10,133	3.35 (average)

Source Director of Agriculture, Jammu and Kashmir State, India

Saffron consists of dried trilobed stigmas of *Crocus sativus*, which is a bulbous perennial with globular corms. The plant is only 15–25 cm tall, native of Southern Europe and cultivated in the Mediterranean countries, in particular, Spain, Austria, France, Greece, besides England, Turkey, India (in Jammu and Kashmir State, an Indian Union territory) and The People's Republic of Iran. It is also grown in the Orient. True Saffron must not be confused with either meadow saffron (*Colchicum autumnale* Linn; Family: Liliaceae) or Safflower or Bastard Saffron (*Carthamus tinctorium*; Family: Compositae), which are occasionally used as adulterants in True Saffron. Saffron is one of the oldest and certainly among the world's most expensive spices. One pound of Saffron consists of about 2,25,000 to 500,000 stigmas which requires piking (by hand) of about 75,000 to 1,25,000 flowers. That gives one a clear idea of the magnitude of human labor involved in the harvest of Saffron. The color of the Saffron is bright yellowish red or orange, and the aroma is very powerful, somewhat bitter, peculiar and exotic, which is mostly due to its stigmas from the fully developed saffron flowers. Saffron, a triploid ($2n = 3x = 254$) reproduces asexually (Lattoo et al. 1997).

Its plant is hysteranthous, with flowers arising directly from corms (Dhar 1991). It is a low-growing plant with an underground globular corm and is cultivated for its large, scented, blue or lavender flowers. The flowers have trifid or trilobed, orange-colored stigmas which, along with the style-tops, yield "saffron of commerce".

Saffron is primarily grown in Pampoosh valley near Bhelgum in Kashmir. The main obstacle in tracking production figures is due to the unusual manner in which plucking of Saffron flowers is done. According to experts, pure Saffron is plucked during early dawn before sunrise. This unusual manner of plucking Saffron flowers is practiced until today in order to maintain its freshness of the filament in the flower and to retain the color of the Saffron filament.

Area, Production and Trade

Table 1.1 gives data on area, production and trade.

Pulwana is the most important district in Kashmir, covering about 80% of the total area and about 78% of total Saffron production in Kashmir. This is followed by Srinagar and Badgam districts, accounting for 12% and 8% of total production, respectively. The highest yield (6.03 kg/ha) is reported from Doda district which is followed by Srinagar district (5.65 kg/ha), compared to the yield in Spain where the

Table 1.2 Trends in area and production of Saffron in India

Year	Area (ha)	Production (tons)	Year	Area (ha)	Production (tons)
1970–71	1628	3.90	1981–82	3022	10.13
1971–72	1644	4.00	1982–83	3699	9.02
1972–73	2024	5.50	1984–85	3841	8.84
1973–74	2028	5.60	1985–86	3566	8.64
1974–75	2109	0.96	1986–87	3655	8.50
1975–76	2127	0.61	1987–88	3796	8.54
1976–77	2191	0.84	1988–89	N.A.	N.A.
1977–78	3185	14.75	1989–90	4000	9.02
1978–79	3255	13.09	1990–91	N.A.	N.A.
1979–80	3405	20.04	1991–92	4001	9.15
1980–81	2615	9.83			

Source Directorate of Agriculture, Srinagar, India/Office of the Financial Commissioner (Statistical Section), Srinagar, India
Spices Board, Government of India. Spices Statistics, 2nd edn. pp. 1–162
N.A. Not Available

yield has been varying from 7.51 kg/ha to as high as 10.14 kg/ha. The present average yield in Spain is 7.61 kg/ha. This calls for yield increase in India.

The available data on saffron production indicate that during 1991–92, 4001 hectares of saffron were cultivated with a total production of 9.15 tons (Table 1.2). The production touched a high of 10.13 tons in 1981–82. Table 1.2 details the break up on acreage and production.

Recent Export Trends from India

In 1995–96, exports touched a level of 7.9 tons valued at INR 133 million (1 US $= INR 75) as against 4.1 tons, valued at INR 13.4 million in 1994–95, almost a tenfold increase in one year. Saffron exporters have revealed that with decline in production, India has continued to hike the price of the produce. Hence, India has been shored up by the international buyers. India mainly exports to Europe, in general, but in particular to Germany, and Japan, but faces stiff competition from Iran, Spain, Greece and Morocco. Countries such as China and Japan produce Saffron for their own domestic consumption. Tables 1.3 and 1.4 summarize the export trends.

World Trade in Saffron

In the world market, The People's Republic of Iran produced around 110,000 kg, followed by Spain, with approximately 30,000 to 35,000 kg, Greece 7000 kg, The People's Republic of China 1500–2000 kg, Japan 1000 to 1500 kg, while Morocco managed to produce 2000–3000 kg. Though Spain is known for exporting double its production, it manages this by procuring Saffron from other countries and re-exports it. In the international market, Spanish brands have an edge, over Indian Saffron, despite the latter having the best quality. This is mainly due to Spain having a good export market strategy. Spain is also known as a consistent supplier of Saffron,

Table 1.3 Trends in Saffron export from India

Country to which exported	1990–91		1991–92		1992–93	
	Qty	Value (INR)	Qty	Value (INR)	Qty	Value (INR)
North America	0.0	0.0	0.0	0.0	0.0	0.0
Australia & Oceanic zone	0.1	141.3	0.1	212.2	0.0	0.0
European Union	5.5	8386.8	9.5	12,768.3	6.4	8,277.1
Rest of Europe	0.0	0.0	0.0	0.0	0.1	141.1
East Asia	0.1	1079.1	0.9	3275.0	0.0	974.8
West Asia	0.1	333.8	1.3	1,592.3	0.1	726.0
Rest of Africa	0.0	0.0	0.0	0.0	0.3	120.1
Total	5.8	10,032.8	11.8	17,974.7	6.7	10,239.1

Source Spices Board 1994. Spices Statistics (Revised) 2nd Edn. pp. 1–162
Notes Qty: Quantity in tons INR= In thousand

Table 1.4 Trends in the annual export of Saffron (Stigma/Stamen) from India

Year	Quantity (Qty)	Total value (INR)	Unit value (INR/kg)
1988–89	9	2.032	5.78
1989–90	13	36,112	7.85
1990–91	6	10,033	1,672.17
1991–92	12	18,095	1,507.92
1992–93	7	10,119	1,445.57

Source DGCI & S, Kolkata. Spices Board 1994. Spices Statistics 2nd Edn. pp. 1–162
Notes Qty: Quantity in tons INR = In thousand

with excellent packing and presentation methodology to the customer. These factors enable Spain to globally command its price. However, the reverse is the case inasmuch as India is concerned (Sandhya 1996).

Spain alone exported 17 tons of Saffron in 1990, as compared to the year 1969–70. Both production and export of Saffron from Spain have declined in recent years, owing to (1) rising labor costs, (2) migration of workers in the industry to cities and (3) competition from synthetic substitutes to Saffron.

Trade in Saffron during 1981–1985 can be seen in Table 1.5, which details production and corresponding import value.

Domestic Fluctuations in Price of Saffron

In Kashmir, India, saffron is mainly traded in three grades, *moongra, lachha* and stems, which are used to blend with pure Saffron. While premium *moongra* and *lachha* are traded more in the domestic and export market only, stems are sold in the local market to blend with the pure Saffron. During December 1996, prices of

Table 1.5 Trends in world trade in Saffron (whole and ground) from 1981–1985

Importer currency	1981		1983		1985		Exporting country
	Quantity (tonnes)	Value	Quantity	Value	Quantity (tonnes)	Value	
Belgium (BF)		18,197	N.A.	15,965		15,213	Spain
Denmark (DK)		136		153		60	Spain
France (FF)	8	12,905	5.0	13,144	4	12,023	Greece, Spain, Iran
Germany (DM)	0.8	2304	1.1	1516	0.3	1198	Greece, Spain, Italy
Italy (Italian Lira)	7.0	5796	7.0	5007	10.0	8033	Greece, Spain
Netherlands(F)	1.0	100	N.A.	103	N.A.	121	Spain
U.K. (Pound)	3.0	197	2.0	241	2.0	324	Spain
USSR®	0.7	309	1.0	303	1.0	301	Iran
Austria (S)	N.A.	N.A.	N.A.	N.A.	N.A.	1235	Greece, Spain
Finland (FmK)	4.0	79	2.0	36	3.0	110	Spain
Norway (NKr)	N.A.	31	N.A.	68	N.A.	77	Not Stated
Spain (Ptas)	1.0	29,363	0.3	5,028	0.1	7414	Greece, Japan, Iran
Sweden (SKr)	2.0	8020	3.0	11,766	2.0	7872	Greece, Spain, India
Switzerland (Sior)	5.0	4137	5.0	1,967	7.0	7520	Greece, Spain, Italy
Algeria (DA)	N.A.	205	N.A.	262	N.A.	N.A.	Spain
Morocco (DH)	N.A.	5	N.A.	1.0	N.A.	2.0	Not Stated
Kuwait (KD)	1.9	576	4.4	6.74	23	1.079	Spain, Gautemala, Iran
Saudi Arabia (SRI's)	46	11,036	30	16,817	39	20.055	Spain, India, Iran
Hong Kong (HKS)	0.4	1537	0.9	1996	3.1	5524	Spain, Greece

Notes Letters in brackets indicate the abbreviations of the local currencies, which according to IMF market rate are equivalent in US dollar. All values are in 000 of the respective currencies
Source International Trade Center (ITC), UNCTAD/GATT, Geneva, 1986. Imports of spices into selected markets during 1981–85. Document presented at the first meeting of the International Spices Group, New Delhi, India

moongra and *lachha* shot up steeply by INR 3000 per kg, to quote at INR 48,000 and INR 38,000 per kg respectively of the two types. The steep rise in prices were attributed to poor yield in the crop season.

Traders revealed that prices during October–November 1996 dipped to INR 40,000/kg against INR 45,000 IN 1995. Prices fell due to panic selling by traders in order to dispose the old harvest. However, by November, new crop came in coupled with the rumor of shortage in harvest; hence, the prices zoomed again back to INR 45,000/kg, and went further up.

There are wide applications of the more expensive Saffron varieties for religious purposes, for the manufacture of "*Zarda*" (the name comes from the original Persian word "*Zard*", meaning yellow, it is flavored tobacco served in Kashmir after a meal) for flavoring and coloring of eatables, in perfumery manufacture, value addition in foodstuffs and preparation of Ayurvedic medicines. Experts reveal that Kashmir farmers of Saffron have not been able to achieve enhanced production of Saffron due to the prevalent political turmoil, who has been on the boil for decades now.

Attempts are now being made to popularize the cultivation of Saffron in the dry temperate hills of India (Kinnaur, Lahaul-Spiti and Pangi-Bharmour in Chamba valley) in Himachal Pradesh (Panwar 1992a, b) and Kumaon hills of Uttar Pradesh (now Uttarakhand) (Munshi 1989).

In view of the decline in acreage, production and export of Spanish Saffron, there is an immense possibility for India to scale up production and enhance export possibilities, through modern methods of cultivation and global marketing strategies. Efforts are already in place in two Indian States, Uttarakhand and Himachal Pradesh.

Crop Management

Soil requirements: Saffron requires a well-drained, sandy to loam soil, free from clay. A medium-grade, well-drained light soil is preferred to rich and very fertile soils, since the latter tend to increase vegetative growth at the expense of reproductive growth. It is necessary to have adequate surface drainage for corm development. Madan et al. (1966) have suggested that Saffron crop has preference for siliceous-argillaceous-ferruginous chalky soils. The corms tend to rot in humid and water-logged soils. This is what happens when the soils contain excessive clay particles, where subsequent to heavy rainfall, waterlogging results, without proper drainage.

An investigation indicated that higher calcium carbonate content in the soil profile of the field of Pampore Valley may be a limiting factor in the successful growth of Saffron there. Further, carbon/nitrogen ratio (C–N ration) in the soil profile does not have any important role to play in the growth and reproduction of Saffron plants. However, these aspects need further investigation.

Climate Requirements

Saffron thrives best in warm subtropical climate. In Spain, it is reported to be grown in dry temperate conditions with annual rainfall rarely exceeding 40 cm. In Kashmir, India, it is found to flourish up to 2140 m above mean sea level. It has also been reported that Saffron prefers cool sunny environment between 1300 and 2300 m above mean sea level. He further noted an elevation of 2000 m above mean sea level to be the best for good flowering; higher altitudes were found to delay flowering, affecting crop yield. Flowering is also found to be delayed when the crop is raised over northern aspects at the same elevation. The best situation for its cultivation may be found on eastern and southern slopes of the hills. Photoperiod exerts a considerable influence on the flowering of Saffron: An optimum period of 11 h of bright sunshine is desirable. Temperature, as well, exerts its influence on flowering, but unusually low temperature coupled with high humidity during the period of flowering will adversely affect it.

Spring rains are considered favorable for boosting production of new corms; a second spell of rainfall at the end of summer or beginning of autumn helps bearing blossom and ultimately increases crop yield. Dry and moderately humid weather conditions at flowering time are ideal; frost or rains during flowering will damage the crop yield. There should be low humidity at the time of the harvest to enable good production of flower crop.

According to Panwar et al. (1995), Saffron can be grown in those cold regions which are situated between 1500 and 2500 m above mean sea level and receive some snow during winters. Empirical studies at Himachal Pradesh Krishi Vigyan Kendra, Research Substation, Sangla, in India, revealed that Saffron can be successfully raised even at those localities which receive heavy snowfall during winters and are situated between 2500 and 2800 m above mean sea level. These high-altitude areas possess dark brown to dark grayish brown, slightly acidic to neutral, gravelly, loamy sandy soils, and experience scalding summers and icy winters; and thus, are suitable for Saffron cultivation.

Land Preparation and Planting

The soil is plowed 25–30 cm deep, harrowed and cleared of all weed growth, penetrating roots and stones and gravels in the soil from March to May. Planting is done in July to mid-August in small beds such that deep drains are laid out after every 6 to 8 rows to enable good drainage. Seed rate is about 8 tons/ha and corms of 2–5 cm diameter or above are selected for planting. The corms are planted between 12 and 18 cm or 15 and 20 cm apart in rows (Madan et al. 1966) and at a distance of 10 cm in the row. The depth of planting is 7.5–10 cm. There is no further cultivation practice until flowering in June, when the soil between the rows is lightly turned over from a depth of 6–7 cm, without disturbing the corms. According to Panwar et al. (1995) Saffron is propagated through corms. Above 500,000 corms (40 quintals, one quintal = 100 kg) of suitable size (2.5 m in diameter), spaced at 20 × 10 cm, between rows and space between corms, respectively, are required for planting one hectare of land. The corms are treated with 5% solution of copper sulfate before planting in July–August. Bigger sized corms produce flowers in October–November in the same year. Mother corms reproduce annually 2–12 cormlets, and some of which also produce flowers in the following year after planting. The flowers are collected manually, daily during morning time and stigmas are separated by hand immediately, after harvest. The corms or plants remain dormant from June to August. Being a perennial crop, its one time planting continues to yield for 8–10 years, old planting are replaced when saffron yields start to decrease, and, its cultivation becomes uneconomical. The crop in subsequent years blooms from the last week of September and continues until end October or, sometimes, until the middle of November. Yields are greatly influenced by wet weather and low temperatures at flowering time. If both, weather and temperature (4–8 °C—during night and 15–20 °C—during day) remain favorable, yields between 2 and 5 kg/ha of dried Saffron can be had, through scientifically managed practices.

Manures and Fertilizers

Srivastava (1973) considered general dressing of 15–22 tons of Farm Yard Manure (FYM) under Chaubhatia (in the Ranikhet district of Uttarakhand State, India) conditions, as basal dressing. In the succeeding years, he recommended a dressing of 50–10–50 kg/ha in the ratio of N–P–K, carried through ammonium sulfate, superphosphate and Muriate of Potash, respectively. Immediately after the harvest, the soil among the rows is dug out to a depth of 15 cm and applied with FYM (Ingram 1969).

Irrigation

Saffron is grown as a rainfed crop in Kashmir, and, therefore, July–September rains are beneficial to the crop. On the contrary, it is grown under irrigated conditions in Spain. Light irrigation during early autumn months helps to accelerate flowering resulting in increased production.

Interculture

In the second year of planting, in June, before the plants resume growth, the soil in between the rows is dug out with the help of a hoe at the right angle to the rows laid out during planting; this facilitates new growth. The soil is broken again before the flowers appear again in October.

Crop Rotation

In Spain, the corms are uprooted after every 4th harvest. This is done usually during the month of May, when corms are well developed. In Kashmir, the crop, once planted, remains in field for 10–12 years (Madan et al. 1966).

Effect of Plant Density on Flower, Yield and Corm Production

Recent investigations on Saffron cultivation in Kashmir revealed that larger-sized corms were produced under low density, whereas at higher densities, small-sized corms were produced. The estimated linear regression was $y = 67.56 - 0.19x$, where y = Saffron yield and x = density of planting. The number of flowers per 100 corms showed significant negative correlation with increasing plant density. These results are based on the investigations conducted in two seasons under ten planting densities, ranging from 49 to 256 corms per square meter (Dhar 1991).

Variation in Floral Characteristics of Saffron in different Kashmir Locations

Recently, Lattoo et al. (1997) conducted a systematic investigation on the question of assessing the temporal variations in Saffron at 4 locations (Chrar-Sharief, Sanatnagar, Pampore and Malabagh in Jammu and Kashmir State, India). Observations on floral characters were recorded from 200 randomly selected flowers at each location. The characters included fresh flower weight, flower size (perianth area), stigma length, fresh and dry stigma weight and Saffron percentage. This investigation indicated that there were appreciable differences in the coefficient of variation percentage for all the floral characters at different locations (Table 1.6). However, the available variability is of no use for fertilization in breeding because of false triploidism. Saffron is basically a clonally propagated crop; hence, its population is expected to be genetically similar even if not totally identical to each other. The variability seems purely due to varied microagroclimate at different locations, since there is not much marked climatic differences prevailing at these locations in Kashmir. The data are summarized below.

Harvesting or Picking of Flowers and Obtaining Saffron

The flowering period starts during middle or late October and lasts only until the first or second week of November. The flower must be picked each morning before the sun gets too hot. They are cleaned during the day, and the style and stigmas are separated from the perianth. These operations require much skilled labor, which coupled with the small yield per acre, accounts for the very high cost of good-quality Saffron. The duration of picking depends upon the time of blooming. The number of flowers and the time of blooming in any year are dependent upon the temperature prevalent in spring and autumn, and, upon the amount of rainfall. A warm spring and log autumn are conducive to early flowering.

Separation of stigmas from the perianth has to be carried out every day, otherwise the flowers wilt. If the flowers are left for a couple of days, operations become very difficult. The collected flowers are picked up individually. The female worker employed for this job holds a flower in her left hand with thumb-nail and forefinger of the hand, she removes the pistils below the perianth, at the same time tearing off the stigmas with the thumb and forefinger of the right hand and depositing these in the containers held for the purpose.

Effects of different levels of X-radiation on vegetative growth and flower yield have been investigated by Singh et al. (1994b) who concluded a dose-dependent decline in vegetative growth, multiplication capacity and flower yield, but the pigment concentration enhanced only slightly.

Effect of Corm Size at Planting

The effect of corm size at planting or the weight of the fresh flower is given in Table 1.7.

Table 1.6 Variation in floral characteristics of Saffron at different geographical locations in Kashmir valley

Flower character	Char Sharief			Srinagar			Pampore			Malabagh		
	Mean	Range	CV (%)	Mean	Range	CV (%)	Mean	Range	CV (%)	Mean	Range	CV (%)
Fresh flower												
Weight (g)	0.30	0.20–0.36	27.46	0.29	0.25–0.36	25.11	0.31	0.18–0.14	24.72	0.37	0.3–0.4	12.1
Flower size												
(cam2)	12.53	7.52–16.6	23.2	12.23	9.44–15.6	19.00	15.9	11 7–18.6	31.2	16.9	13.119	713.3
Stigma												
Length (cm)	2.92	2.63–3.25	8.41	2.46	1.78–3.0	17.3	2.16	2.33–37	11.4	3.7	2.8–5.1	6.42
Fresh												
Weight (g)	0.03	0.01–0.03	19.1	0.02	0.01–0.05	24.4	0.27	0.01–0.46	51.03	0.03	0.02–0517.4	
Dry stigma												
Weight (g)	0.04	0.02–0.01	25.7	0.004	0.004–0.002	27.9	0.01	0.002–0.001	36.2	0.005	03–0.08	23.23
Saffron (%)	8.24	5.9–11.1	17.5	6.1	4.43–8.21	14.2	8.46	6.03–14.1	32.1	7.5	7.72–11.7	11.66

Notes Saffron Percentage = $\frac{\text{Dry stigma weight} \times 100}{\text{Fresh Flower Weight}}$

Source Lattoo et al. (1997)

Table 1.7 Effect of size or weight of mother corm on fresh flower yield in Saffron

Corm diameter at planting (cm)	Corm weight (g)	Fresh flower yield (g)[a]			
		1990	1991	1992	1993
1.0	3.2	0.86 ± 0.026	3.60 ± 0.76	43.20 ± 8.60	34.60 ± 5.36
1.5	4.5	9.60 ± 1.73	34.40 ± 4.76	74.60 ± 7.50	63.80 ± 8.42
2.0	6.1	21.20 ± 2.93	51.20 ± 6.38	76.40 ± 9.32	60.40 ± 8.54
2.5	8.5	24.40 ± 3.01	60.00 ± 5.10	84.20 ± 8.46	68.40 ± 9.32
3.0	12.4	27.60 ± 3.26	63.40 ± 6.34	75.40 ± 12.34	58.60 ± 8.04
3.5	18.2	30.70 ± 4.17	70.30 ± 7.30	64.60 ± 10.38	53.90 ± 9.56

Notes [a] Yield data from 200 corms ± Standard Deviation
Source Singh et al. (1994b)

Yield

The number of flowers and the time of blooming are dependent on the temperature prevalent in spring and autumn and the amount of rainfall. A warm spring and a long autumn are conducive to early flowering (Madan et al. 1966). In the first year, the yield of Saffron is negligible, but it increases gradually and during the third year of flowering, an annual yield of 1800–2500 g of Saffron is obtained from one hectare of land (Srivastava 1964). Other investigators have reported maximum yield up to 6 kg/ha. It may be of interest to note that about 147,000–166,000 flowers give 5 kg of fresh stigmas that go to make a kilogram of pure dry Saffron. Ingram (1969) reported that an yield of 10 kg Saffron/ha was produced from about a million of blossoms. Proper irrigation and fertilizer management lead to such good Saffron yields.

Cost of Cultivation of Saffron: Cost and Returns Structure

The establishment cost of Saffron cultivation, on average, came to INR 70,260/ha (1 US$ = INR 75, INR = Indian Rupee) of which the treated seed corms formed the major component, that, is 69.3% as shown in Table 1.8. This was followed by interest on capital which came to about 11.93%, manures and fertilizers 8.67%, field operations, such as, planting seed corms, application of manures and fertilizers, weeding, irrigation, strip making, picking, separating and drying of Saffron came to about 95.35%. Unforeseen expenditure came to about 2.84%, and rental value of land is 1.71%. A farmer owning his own land would invest 98.29% of the total establishment cost for undertaking Saffron cultivation in India. The total maintenance cost came to INR 126,970 (Table 1.8), over a period of 9 years. This varied from the second year after planting from INR 10,436 to INR 18,872/ha in the tenth year of cultivation. The maintenance cost, however, remained stable from the 5th year onward until the 9th year. Details are given in Table 1.8.

The maintenance cost in the 10th year also included expenses on the digging out of the Saffron corms and their cleaning, etc., an operation taken up in July of the following year. Picking, separation and drying are very important operations in the

Table 1.8 Cost and returns structure of Saffron cultivation in India

Cost-returns	Establishment period					Maintenance period				
	1st Year	2nd Year	3rd year	4th year	5th year	6th year	7th year	8th year	9th year	10th year
A. Production cost										
1. Field Preparation	800									
2. Manures and fertilizers	6092	3092	3092	3092	3092	3092	3092	3092	3092	3092
3. Application of Manures and Fertilizers	165	138	138	138	138	138	138	138	138	138
4. Seed corms, their treatment and planting	49,388									
5. Strip making	83	83	83	83	83	83	83	83	83	83
6. Interculture operations	1375	1375	1375	1375	1375	1375	1375	1375	1375	1375
7. Irrigation	330	330	330	330	330	330	330	330	330	330
8. Picking, Separation and Drying	412	1100	2062	2475	4950	4950	4950	4950	4950	4950
9. Top dressing	82	82	82	82	82	82	82	82	82	82
10. Digging out Saffron seed corms	–	–	–	–	–	–	–	–	–	3550
11. Rental land value	1200	1200	1200	1200	1200	1200	1200	1200	1200	1200
12. Unforeseen Expenditure	2000	2000	2000	2000	2000	2000	2000	2000	2000	2000
Interest on Capital invested @14% excluding Unforeseen expenditure	8383	1036	1171	1221	1575	1575	1575	1575	1575	2072
Total expenditure	70,260	10,436	11,533	12,004	14,825	14,825	14,825	14,825	14,825	14,825
B. Returns										
14. a. Saffron (kg)	0.114	0.221	0.448	0.787	1.142	2.765	1.650	5.155	1.040	0.775
b. Saffron seed	–	–	–	–	–	–	–	–	–	100
15. Gross Returns in INR: a. From saffron sold@INR 21,000/kg	2394	4641	9408	16,527	23,982	58,065	34,650	108,255	21,840	16,275

(continued)

Table 1.8 (continued)

Cost-returns	Establishment period					Maintenance period				
	1st Year	2nd Year	3rd year	4th year	5th year	6th year	7th year	8th year	9th year	10th year
b. From Saffron seed corms sold@ INR 1200/g	–	–	–	–	–	–	–	–	–	120,000
Total returns (INR)	2394	4641	9408	16,527	23,982	58,065	34,650	108,255	21,840	16,275
Net returns										
Cost of treated										
Seed corms	−67,866	−5795	−2125	4523	9157	43,240	19,825	93,430	7015	117,403
				48,828						
Cost of planting	510									
Total cost INR	49,338									

Table 1.9 Economic viability of Saffron cultivation

Techniques employed	Estimates		
	10 years	9 years	8 Years
NPW—net present worth (INR/ha)	29,982	33,926	36,885
IRR—internal rate of return (%)	21.95	22.97	23.89
BCR—benefit–cost ratio	1.26	1.30	1.34
PBP—pay back period (years)	4	4	4
Average annual income (INR/ha)	21,881	24,151	26,293

Source Pawar et al. (1995)

cultivation of Saffron, and these together account for 27.83% of the total maintenance cost. Similarly, investment on manures and fertilizers, their application, strip making, weeding and irrigation; unforeseen expenditure; interest on capital invested and rental value of land amount to INR 21.92; 14.23; 14.18; 10.54 and 8.50% respectively. Digging out Saffron corms accounted for 2.79% of the total maintenance cost. The crop reached the break-even point in the 4th year since planting. The yields of Saffron increased from 0.114 to 5.155 kg/ha from the year of planting to the 8th year of its perennial life. Yields, however, start declining from the 9th year since planting. Per hectare gross returns increased from INR 2394 in the first year of planting to INR 136,275 in the 10th year since planting. The value of seed corms (after completion of 10 years since planting) accounted for the major part of the gross income, which was, 88.08%. Net returns per hectare were negative in the first three years since planting, obviously, due to high expenditure on seed corms and lower yield of Saffron. Munshi (1989), however, reported negative returns during the first year of planting only. From the 4th year since planting, net returns increased from INR 4523 to INR 117,403 in the 10th year since planting. The remunerative price of seed corms in the 10th year since panting mainly contributed to increase in net returns.

Economic Viability of Saffron Cultivation

Table 1.9 summarizes data on the economic viability of Saffron cultivation.

The data in the above table represent different parameters of economic viability. The net present worth was observed to be INR 29,982/ha. The average annual income per hectare stood at INR 21,881. The BCR, IRR and PBP were found to be 1.26%, 21.95% and 4 years respectively. Estimates of economic viability such as BCR, which was greater than unity and IRR which was also much higher than the considered rate of interest (14%), suggested a sound case for scientific cultivation of Saffron in high-altitude areas. Since the yield of Saffron started declining from the 9th year onward, an attempt was, therefore, made to arrive at the optimal period of its cultivation. Two alternative periods of 9 years and 8 years cycle were considered to evaluate economic viability of the crop cycle, which revealed a considerable improvement in estimated parameters of NPW, BCR and IRR (Table 1.9) over 10 years cycle. Based on these results, a crop cycle of 8 years is suggested for optimal Saffron returns.

Saffron is a winter active and summer dormant (June to August) crop, and, therefore, provides opportunities to grow other short-duration vegetables like French bean and peas, besides, black caraway as inter or companion crops. Data on such combinations were, however, not available, which otherwise would further improve per hectare returns from Saffron and hence its economic viability.

The findings of this investigation have established that cultivation of Saffron on a large scale requires huge initial investment, 70% of which is spent on obtaining seed corms alone. For lowering down this cost, it is advisable to start Saffron cultivation with smaller plots, and corms multiplied in these small plots can be used as base seed material for increasing area in the subsequent years. As mentioned earlier, the production of Saffron keeps increasing until 8th year since planting, when production declines; as such, to realize maximum gains, the saffron crop cycle should be restricted to 9 years. In the 10th year since planting, corms should be dug out and used as seed corms for the subsequent planting. Similarly, cultivation should be encouraged through agricultural cooperative societies in compact areas. This, on one hand, would facilitate dissemination and adoption of scientific agricultural technology among the growers, and, on the other, will pave the way for creating better marketing avenues for the harvested produce.

Pests and Diseases

The crops in the State of Jammu and Kashmir, in particular, in Kashmir, are free of pest and disease attack, though rats create problems on digging up the underground corms and damage them.

Recent Investigations on Crop Management of Saffron in the People's Republic of Iran

The first ever seminar on Saffron was organized by the Iranian Research Organization for Science and Technology, Kharrasson Branch on 8–9 November 1988, at Ghaen. Some of the highlights of the conference on Saffron, which is the original native of Iran, especially with regard to crop management, are given below:

Breeding/Cytogenetic Aspects

While discussing cytogenetic aspects relating to breeding, a rare existence of 3-branched style of Saffron has been reported. It is not stated whether each branch is trilobed, as in the Indian Saffron. If so, its multiplication on a commercial scale to increase Saffron yield would be very welcome, as it can increase yield by three-fold. This will lead to huge commercial benefits.

Effect of Chemical Fertilizers

The results have shown that nitrogen fertilizer has the greatest effect on increasing flower yield. Addition of phosphatic and potassic fertilizers led to further increase, but these increases were not statistically significant.

In soils with 1% organic matter, there was no positive effect through the application of FYM (Farm Yard Manure) or other organic manures.

The results of an 8-year-old investigation showed that 100 kg N/ha had the best effect on Saffron in increasing flower yield. However, higher rates of Urea fertilizer decreased flower yield.

In some cases, application of 30 tons of cow manure plus 50 kg/ha of ammonium phosphate increased flower yield. However, this increase was neither statistically significant nor commercially economic.

Field trials at Esfahan, the second biggest city next to Tehran, which is the capital of the country, revealed that average yield of dry cured Saffron was 6150 g/ha which would fetch about 95,000 Iranian Riyals/ha as net profit to the farmer per annum.

Cultural Practices

Planting of corms is done prior to broadcasting of the fertilizer/manure. This is followed by field irrigation. In subsequent years, the saffron fields receive a shallow till (before irrigation) in the beginning of November. All cultural practices from planting to harvesting are done manually, by hand.

Average Saffron Yield

The saffron yield is low in the first year of planting, but increase in subsequent years and peaks in the 5th and 6th years since planting. Average annual flower yield in a 10-year period is 561 kg. Every kilogram of flower yields only 10.96 g of Saffron.

Pests of Saffron

The most significant pest of the crop in India is saffron bulb larva, which belongs to the Crambidae family.

Weeds and Weed Control

Important weeds of Saffron: They include poa, pigweed, lambsquarters, wild barley, wild radish and descorynia. Weeds are mainly controlled mechanically. Observation plot during 1983 demonstrated that Fusilade and Betanal are effective weedicides to control weed menace in Saffron. They control grassy and broad-leaved weeds.

Weed control is an effective step in improving Saffron yield quantitatively and qualitatively. The first step in a weed control program is the identification of the weeds. In this investigation, a survey was undertaken to map the weeds of the Saffron fields in Khorasan, an important region for growing Saffron in Iran, and, perhaps, in the world. Several Saffron farms in Gonabad, Begestan, Ghaen and Birgand were selected accordingly considering the area of Saffron cultivation. Random samples of weeds in one square meter were sampled during spring and summer of 1988–89 from the selected fields. The weeds were initially counted and evaluated visually for their genera. The results indicated that over 200 species of weeds existed within the Saffron fields, with about 20 species dominating. The most significant weeds during spring season were the following: *Cadaria draba, Holosteum umbellatum, Hordeum spotaneum;* during summer season, the following were the important weeds: *Cardara draba, Alhagi camelorom and Hordeum spontaneum.* Prior to summer, the deep rooted, drought tolerant, perennial weeds gradually replaced the ones found in spring season.

Utilization of Saffron leaves as feed For Sheep and Goats

A comparative investigation of Saffron leaf digestibility was done, which indicated that the average ad lib of dry and fresh organic matter of Saffron were respectively 600 and 526 g/day for sheep and 273 and 240 g/day for goats of Khorasan.

Digestability of dry matter, organic matter and other total nutrients for sheep was 59.54%, 161.54%, 56.01% and 63.82%, 64.38% and 58.2% for goats. Sheep and goats digested 55.45% and 58.21% Saffron raw energy.

Ad lib of dry and fresh organic matter and digestibility of dry matter for sheep and goats were significantly different both at 5% and 1% confidence level.

The ad lib of dry matter for sheep was more than for goats. And, digestibility of dry matter for sheep was less.

Solar Drying of Saffron

The first step in preparation of Saffron after harvest is drying. In Iran, this process is done using ancient methods which result in poor quality produce. In order to correct this situation, the Iran Industrial and Research Organization (IIRO) designed a solar dryer which improves crop quality and maintains its purity. Since many valuable volatile compounds of Saffron may be lost, if improperly dried, the design of the solar drier for Saffron must incorporate knowledge of the chemical and physical characteristics of Saffron. Selective drying methods facilitate maintenance of Saffron quality. To find necessary data, Saffron was dried in different conditions of temperature, humidity, and wind speed, and, the kind and amount of lost volatile compounds identified.

Post-harvest Technology

The value of Saffron depends mainly on its color, flavor and overall good quality, which in turn depends on the method by which the stigmas are dried. The techniques followed in Kashmir and Spain are described below:

Sun Drying of Saffron in Kashmir

In Kashmir, the stigmas picked from the flowers and sun-dried to 8–10 or 10–12% moisture are arranged and constitute the first grade ("*Shahi Saffron*"). The flowers are dried in the sun for 3–5 days, then slightly beaten with sticks and passed through coarse sieves. The material which passes through is thrown into water, those parts of the flowers which float are discarded and those which sink to the bottom are collected and further dried, constituting the second grade ("*Mongra Saffron").* The discarded parts of the flowers are again subjected to the beating process, and the process of throwing the entire pounded mass in water is repeated. The product which sinks is collected, which is very much inferior in value, thus constituting the 3rd grade saffron ("*Lachha Saffron*").

Alternatively during bumper harvest, the whole flower harvest is dried in the sun and stigmas are picked later by hand from the dried flowers. In another process, prevalent in Kishtwar region, the entire style with stigma is bound into a small bundle or "*Gucchi*" soon after harvest of flowers, and dried in the sun.

Table 1.10 Physico-chemical characteristics of stigmas at various stages of flower development

Development Stages of flowers	Average weight of stigmas (g)[a] ($n = 15$)	Average length and diameter of stigmas (mm)[a] ($n = 15$)		Crocin content in fresh stigmas (mfb)[b]/g/kg
		Length	Diameter	
Bud (first day)	0.026	28.0	3.0	165.4
Bud opening (third day)	0.032	31.5	3.6	167.2
Full bloom (fifth day)	0.037	35.0	4.0	170.7

Notes [a] Calculated for three consecutive years from different locations
[b] *mfb* Moisture-free-basis
Source Raina et al. (1996)

In Spain, where the process is called "*Toasting*", the stigmas are placed in sieves in layers 2–3 cm thick over an almost spent fire. The sieves are placed 15 cm above the fire and by stacking them and changing their order and position, the product is carefully dried. The process, in addition, may utilize special stoves for the purpose. The saffron has to be kept protected from dampness and light because light bleaches it into dull yellow color.

The yields of dried Saffron (2–2.5 kg/ha) both in Kashmir and Jammu, in India, are much lower than those obtained in other countries, where yields of 3.5–8 kg/ha are obtained. About 5 kg of fresh stigmas yield 1 kg of dry Saffron, which is about 20% of the original weight.

The final product as sold in the *bazars* (Indian markets) is a loosely matted mass of dark, reddish-brown flattened stigmas, with characteristic aromatic odor and bitter in taste. When fresh, they are glossy and unctuous to touch, but after keeping, they become dull and brittle.

In order to obtain a high-quality product, it could be desirable to check the process of hydrolysis of crocin and allied pigments during processing. A detailed study of current techniques (Sampathu et al. 1984) confirmed that the traditional methods of sun drying in India takes about 27–53 h. Such long drying period might be responsible for enzymatic degradation of crocin. Therefore, it is logical that a shorter drying period and a controlled temperature may give the desired high-quality Saffron.

The quantitative distribution of pigments in stigmas at various stages of flower development was estimated by using UV spectroscopic method (data in Table 1.10). The stigmas from flower bud to full bloom stage were separated and the pigments in the fresh matter estimated. The pigment composition estimated from the spectra was consistent, and overall pigment concentration was 16.5–17.0% on a moisture-free basis (Tables 1.10 and 1.11).

However, the increased weight of stigmas at full bloom was associated with more pigment per flower. Such a high percentage of crocin pigments in fresh stigmas confirmed that the germplasm cultivated in India is of an intrinsically high quality and indicated the deficiencies in post-harvest processing were responsible for the production of poor quality Saffron. India predominantly dries Saffron in the sun, whereas the top quality source—Spain—uses artificial drying (Raina et al. 1996).

Table 1.11 Contents of crocin pigments in Saffron after various artificial processing conditions

Drying protocol	Crocin (g/kg)a	Drying time (h)[a]
Shade drying (4–18 °C)	110.0	53.0
Sun drying (15–21 °C)	127.0	28.0
Solar drying	153.0	6.5
Inlet air temperature (35–49 °C)		
Outlet air temperature (26–40 °C)		
Electric oven drying at 0 °C		
20 °C	124.0	22.0
40 °C	154.6	5.3
50 °C	148.5	3.8
65 °C	134.0	3.0
80 °C	115.5	2.4
Cross-flow drying at 0 °C		
20 °C	129.4	25.0
40 °C	156.4	7.0
50 °C	150.4	4.5
Vacuum oven drying (at 0 °C) and reduced pressure (40 mm)		
40 °C	170.5	5.0
50 °C	161.1	3.6
65 °C	144.7	3.0
Dehumified drying (at 0 °C)		
40 °C	156.7	4.6

Notes [a] The investigations were conducted during three years continuously on samples (four each) collected from two locations in Kashmir, India
Source Raina et al. (1996)

Effect of Different Processing (Different Drying and Dehydration Techniques)

Table 1.12 presents data on the main flavoring components of Saffron flavoring principles as affected by the different methods of processing (sun drying, solar drying, cross-flow dehydration, vacuum shelf dehydration, etc.) at different temperatures of dehydration (20–50 °C).

Dehydration Characteristics of Saffron

Depending on the temperature and tray load, the drying characteristics responded differently. The total drying time needed an optimum tray load of 1 kg/square meter for drying saffron by different methods varied between 2.4 h (oven drying at 80 °C) and 53 h (in shade). The best drying was achieved in a solar cabinet for 6.3 h and in an oven at 40 ± 5 °C (4.0–5.3 h). The traditional method of sun drying took 22 h and yielded a poor quality product. The solar and oven dryers were most effective in reducing initial moisture content by 50% during the first hour of drying.

Table 1.12 Effect of different methods of drying and dehydration on main flavor components in Saffron

Saffron drying techniques employed	Volatiles (g/kg)	Safranal (in volatile oil, g/kg)	4-β-hydroxy Safranal oil g/kg)	Other (in volatile compounds
Fresh samples	Trace	–	–	–
Shade-dried, sun-dried, solar-dried, electric oven dried (at 20, 40, 50, 65 °C) or dehumidified dried	6.0	550–680	140–220	Pattern similar to that reported in literature
Vacuum or cross-flow dried	5.0	200–350	500–700	Two major peaks in GLC; identified as Safranal and 4-hydroxy Safranal

Source Raina et al. (1996)

Sensory Characteristics

Flavor profiles of all the samples of Saffron were obtained with 14 panelists following standard procedures. Flavor profile analysis at fixed dilution was carried out. The results showed that solar-dried and oven-dried samples at 40 ± 5 °C were most intense, with dominant floral, sweet and spicy notes and few harsh and acrid notes; they were adjudged to be the best. Vacuum oven and cross-flow dried samples were most unbalanced with poor initial notes and dominating herbal and barky end notes. Shade-dried samples, though low on intensity of initial notes, had a pleasant flavor profile. Further, these results are consistent with the chemical composition of the samples.

Development of a Solar Drier for Saffron

A small solar drier suitable for the daily output by an individual Saffron grower/farmer has been designed and fabricated, using locally available materials, like wood, corrugated metal sheets, and glass panes, etc., by the Regional Research Laboratory (RRL), Jammu, in Jammu and Kashmir State, India, affiliated to the Council of Scientific and Industrial Research, (CSIR), New Delhi, India. The cost of this solar drier is about US $ 25 on a price parity as prevalent in 1995, when the drier was first fabricated. Since then, the cost has greatly increased. The unit is provided with a single tray of 80 cm × 540 cm, fitted with a coarse silk cloth.

Storage Investigations on Saffron samples under Ambient Conditions

The coloring power of dried Saffron was evaluated during 36 months of storage at regular time intervals. The results showed that by storing saffron in 100-gauge polyethylene bags at ambient temperature (15–35 °C) in dark over a one-year period, the loss of crocin varied from 56.9 to 70.2%, depending on the moisture content (10–12%) (Alonoso et al. 1990). However, the trend declined during subsequent storage. The results confirmed that under Indian temperature conditions, the admissible moisture conditions (as per ISO standards) were sufficient to catalyze degradation (hydrolysis) of crocin into crocetin, thus lowering coloring power of Saffron. The samples with initial moisture content of 5% deteriorated more slowly. The coloring capacity tended to be inversely related to moisture content. Packaging and storing Saffron in polystyrene box cover-wrapped with MXXT cellophane film checked degradation of pigment and flavor components.

Concluding Remarks on Post-harvest Processing and Storage

The above investigations by Raina et al. (1996) have confirmed that the Indian germplasm of *Crocus sativus* is of high quality and the flowers should be harvested at the full bloom stage to obtain high yields of Saffron. It was also concluded that the content of important quality substances differed after different types of artificial drying. The relationship between post-harvest processing temperature and quality of finished products has been established. Processing at 40 ± 5 °C (solar dryer/oven dryer) at a tray load of 1 kg/square meter is optimum for the production of high-quality Saffron and results in large time saving over traditional sun drying. Adoption of artificial drying by either of these two methods should ensure that Indian Saffron would conform to ISO specifications. Storage investigations also revealed that packaging of Saffron with an initial moisture content of about 5% in polystyrene box over wrapped with MXXT cellophane increased shelf-life of the finished material.

Summary

Quite significant variations have been observed in the concentration of pigments (10–10.7%, moisture-free-basis) and flavor components during different post-harvest processing conditions of Saffron (stigmas of *Crocus sativus*). The concentration of crocin pigments was the highest (15–17%) in the Saffron samples dried between 35 and 50 °C (see Table 1.13), either in a solar dryer or in an oven dryer, and, thus also resulted in considerable reduction of normal drying time. Under these conditions, the main flavoring component Safranal was at its peak value of 60% in the oil in almost all of the samples, except the vacuum-oven-dried samples, which interestingly contained 4-β—hydroxysafranal in major amounts. Investigations indicated that 4-β-hydroxysafranal may be an intermediate in the formation of safranal. It was observed that prolonged storage affected pigments and flavor concentration to a great extent, but proper packaging and storage with 5% moisture in the Saffron samples reduced deterioration in quality, thereby increasing its shelf-life.

Composition

The average composition of commercial Saffron is given in Table 1.14.

Yield of essential oil and fixed oil as high as 1.37% and 13%, respectively, has been recorded. The ash is rich in potassium and phosphorus and contains traces of

Table 1.13 Changes in the Crocin pigment contents (g/kg) and volatiles in Saffron[a] dried/dehydrated by different techniques, stored at ambient temperature (15–35 °C)

Saffron drying technique	Storage period (months) at 15–35 °C							Loss of Volatiles after12months storage at ambient temperature (15–35 °C)
	0	3	6	9	12	18	24	
Sun drying	127	111.7	94.8	63.2	50.6	44.3	35.5	(1) 50% Volatile oil lost resulting in low flavor intensity
Solar drying	153	134.2	114.3	76.4	92.0	46.0	30.7	(2) GC pattern was stable
Electric-oven drying (40 °C)	154.6	136.0	116.0	76.9	62.0	52.6	31.0	
Vacuum-oven drying (at 40 °C)	170.5	145.0	119.0	86.2	63.3	51.0	34.2	(1) Oil composition and percentage changed significantly
Cross-flow Dehydration (at 40 °C)	156.4	132.5	114.2	78.0	54.7	47.0	31.3	(2) 4-β-hydroxy Safranal declined from 60–20%

Notes [a] Stored in the dark at ambient temperature with 10–12% initial moisture
Source Raina et al. (1996)

Table 1.14 Average composition of Saffron

Property	Content (%)
Water/moisture	15.6
Starch and sugars	13.35
Essential oil	0.6
Fixed oil	5.63
Total N-free extract	43.64
Crude fiber	4.48
Ash	4.27

boron. Kashmir Saffron contains 8.5–10.2% moisture and 5.9–13.3% total ash. The ISI has laid down standards for Saffron and so also the AGMARK authorities.

Adulteration

Because of its high cost, Saffron is frequently adulterated with styles, anthers and parts of corolla of Saffron. Exhausted Saffron flowers and floral parts of some Compositae like *Calendula* spp. and *Carthamus tinctorius*, corn silk and various materials colored with coal-tar dyes are also used as adulterants. Water, oil or glycerin is added to increase its weight. Cake-saffron of commerce often contains safflower florets with adhesive sugary substances. Experienced Saffron tasters of Kashmir can correctly estimate the degree of adulteration.

Packaging and Storage

Care has to be taken to prevent deterioration of the quality of Saffron while in storage. For this, the dry stigmas are stored in cloth-bags or butter paper and packed in closed tins or in air-tight wooden boxes. It is found that packing of Saffron in polyethylene bags and keeping in air-tight containers (tins) maintains the quality of Saffron over extended periods of time. However, it is necessary to dry Saffron properly before storage. Saffron, when stored for more than one year, sometimes, turns dark in color and loses its characteristic taste and aroma.

Marketing

Method of marketing Saffron in Kashmir continues to be quite traditional. The produce is generally sold by the growers/farmers themselves to itinerary merchants or directly to wholesale merchants. Sometimes the producer sells the produce directly to the wholesale merchant and barter other family requirements. There are no regulated markets or Saffron in Kashmir, in India.

Pricing

According to the ITC, the average prevalent export price of Saffron was US $ 16454/28 tons of Saffron exported or about US $60/kg or INR 9000/kg. Today, a kg of Indian saffron costs INR 1,20,000 (77 INR = 1 US$). Saffron price has skyrocketed after the Afghan war. With decline in saffron production in Spain, India, which holds the second position in world market in saffron, stands to greatly benefit if the government pays more attention to this costliest spice in the world.

Processed Products

Saffron being the costliest spice in the world, not much attention has been paid to the manufacture of any processed product from Saffron. Only in 1996, Raina et al. (1996) have reviewed and also conducted systematic investigations in pigments and volatiles (aroma deciding factors) of Saffron during processing and storage under ambient conditions (1535 °C). No value-added products have emerged from saffron. Obviously, there is considerable scope for more intensive research and development effort on the modernization of post-harvest processing and efficient and attractive packaging.

Uses

As Flavorant and Colorant for Foods

Saffron is famous for its extraordinary medicinal (pharmaceutical), flavoring (nutraceutical) and coloring properties. It is used overseas in exotic dishes, in particular, Spanish rice specialities and French fish preparations. It is also used in fine bread making in many countries, including Scandinavia and the Balkans. The principal coloring agent of saffron is glycoside crocin; the bitter substance is the glucoside picrocrocin.

Medicinal/Pharmaceutical Properties and Uses

Saffron is credited with various medicinal/pharmaceutical properties. It is used occasionally in exanthematous diseases, which promote eruptions, such as viral infections like measles in children or chicken pox, which produce eruptions on skin. It is used to control fever, melancholia and enlargement of liver and spleen. It has also stimulant and stomachic properties and is considered to be a remedy for catarrhal affliction in children. In modern pharmacopeia, it is employed only to color other medicines or as a cordial adjunct. Saffron has been employed as an abortifacient and successfully used in several fatal cases. Saffron bulbs are toxic to animals, and stigmas in overdose act as a narcotic.

Saffron is an important ingredient of the *Ayurvedic* and *Unani* systems (Indian indigenous systems of medicine) of medicine. It is popularly known as a stimulant and is warm and dry in action and is helpful in urinary, digestive and uterine problems. It is mixed with other drugs to help in normal menstruation. It is soaked overnight in water and is used with honey, which helps patients suffering from uterine problems to pass urine without difficulty.

In homeopathy, the mother-tincture, prepared from Saffron, and its suitable dilutions are useful in about a dozen of human disorders related to mind, head, eyes, nose, throat, stomach, male and female organs, anus, respiratory and sleep disorders.

The volatile oil contained in Saffron is used as an external application in uterine problems, such as uterine sores. When pounded with ghee (locally made, *desi ghee,* Indian butter*)*, it is reported to be very effective in regulating blood sugar, hence, in diabetic patients. It is also reported to give strength to the brain—a brain activizer—but only when used in large doses.

Tulasi

Botanical name: *Ocimum sanctum*

Family: Labiatae

How many of us really know the value of Tulasi? All that we think is that it is a sacred plant and used only for worship in the Hindu temples and at home. Almost all devout Indians cannot think beyond this. But Tulasi is an exceptionally valuable plant inasmuch as human nutrition is concerned, when it comes to its seeds. Let us examine the details below.

It is a cost-effective substitute for eggs in cake baking, breads and biscuits, just a spoonful of Tulasi seeds is enough for a fluffy cake batter. Tulasi seeds are called *Sabja* in northern India and are used in, put in tea, to treat sore throat and running nose. Tulasi herb comes under the genus *Ocimum.* Its botanical name is *Ocimum sanctum* which belongs to the family Labiatae. The word "sanctum" means sacred. The genus *Ocimum* has over 150 species spread across the world. In Tulasi, we have both Rama Tulasi and Krishna Tulasi, which are sacred for use in temples and homes.

Ocimum basilicum (also from the genus *Ocimum* like Tulasi) is a prolific seed producer. The seeds have become very popular among health conscious people. The small black seeds (see picture enclosed) become mucilaginous (gain a gelatinous consistency) when soaked in water and consumed are said to help reduce body weight, help patients suffering from constipation and acidity and also manage diabetes. It is rich in in dietary fiber such as the polysaccharides glucomannan and xylan which make one feel satiated after a long time. An investigation published in the science journal Bioactive Carbohydrates and Dietary Fibre of May 2021 issue establishes that the fiber accounts for 98.50% of this mucilage.

The hydrated seeds do not have a strong flavor or taste; they can be easily added to smoothies (sweet milk/fruit based drinks), fruit juices and milkshakes. They have been used as a replacement for fats in sponge cakes by researchers of the department of Food and Nutrition, Korea University in Seoul. In a 2017 investigation reported in the Italian Journal of Food Science, the researchers report that 1 g of seeds can replace 15 g of butter.

Nonhydrated seeds are also edible. In 2020, an investigation reported in the science journal Current Research in Nutrition and Food Science, researchers from the Department of Nutrition and Dietetics at Periyar University, Salem, in Tamil Nadu, India, showed that using roasted Tulasi seeds in *idli* (a very popular south Indian breakfast fare) batter made the idlis more healthier and suitable for diabetic patients.

The seeds are used in traditional recipes too. A nutritious *pak* or burfi or mysore pak (both popular sweet dishes of India) is made by turning the seeds into flour and mixing them with ingredients such as almond, *khoya*, Bengal gram flour, black pepper, crystal sugar and ghee. This sweet can be eaten every morning on an empty stomach with milk for good nutrition.

Water Conserving Plant

Tulasi is water efficient. The increasing popularity of Tulasi, for its stems, leaves and seeds, more in North India than in Kerala, has attracted many farmers to take to its cultivation on a large scale. It can be successfully grown in water-starved (rained) areas. In fact, water stress, along with high and low temperatures, increases the essential oil content of the Tulasi seeds.

The Tulasi seeds are not just used in cooking. They yield an essential oil which has antimicrobial properties and inhibits the growth of yeasts and fungi on the skin. For instance, the essential oil from Tulasi inhibits the germ tube formation in the yeast *Candida albicans* which causes infections in humans and prevents their growth, according to an investigation published in the science journal Industrial Crops and Products.

In recent years, there have also been extensive investigations on the Tulasi mucilage to examine if the same could help develop materials which can be used in personal care products, cosmetics, medicines and in agriculture.

Since Tulasi grows profusely, the Central Institute of Medicinal and Aromatic Plants in Lucknow, Uttar Pradesh, has developed hybrids which give diverse fragrances and flavors such as that of *paan* (beeda), lavender, lemon and cardamom. Many of the *Ocimum* species already have flavors similar to some of the other herbs. For example, Thai Tulasi has an anise-like fragrance, while African basil has a strong camphor-like smell. These hybrids serve as cheap substitutes in the food industry, thus paving the way as good nutraceuticals.

References

Alonoso GL, Varon R, Gomex R, Navarroo F, Salinas MR (1990) Auto-oxidation in saffron at 40 °C and 75% relative humidity. J Food Sci 55:595–596

Dhar AK (1991) Studies on saffron in Kashmir-1V. Variation in corm size and its effect on corm production and flowering. Indian Perfumer 35(3):173–176

Ingram JS (1969) Saffron. Trop Sci 11:177–184

Latoo SK, Dhar RS, Dhar AK (1997) Temporal variation in saffron. J Spices Aromatic Crops 6(1):57

Madan CI, Kapure BM, Gupta US (1966) Saffron. Econ Bot 20(4):77–85

Munshi AM (1989) Economic analysis of saffron under rainfed conditions of Kashmir. Agric Situation India 44(5):379–381

Panwar KS, Saroch K, Vashist GD (1995) Potential and prospects of saffron in temperate hills of Himachal Pradesh. Agric Situation India 49(4):13–16

Panwar KS (1992a) Uchch parvatiya shuskh shitoshan kashetar—"Kesar Hi Kesar". Phal Phool (ICAR) 15(1): 11–12, 16

Panwar PS (1992b) Saffron—an ideal cash crop for high hills. Farmer Parliament 27(12):2, 25

Sampathu SR, Shivashankar S, Lewis YS (1984) *Crocus sativus* Linn. cultivation, processing, chemistry and standardization. Crit Rev Fd Sci Nutr 20(2):123–152

Sandhya DM (1996). Saffron. Economic Times, New Delhi, India

Singh C, Ram G, Kaul BL (1994a) Saffron studies in Kishtwar. Effect of corm size at planting time on corm production and flower yield in *Crocus sativus*. Indian Perfumer 38(3):82–84

Srivastava RP (1964) Cultivation of saffron in India. Fertl News 8(9):7–11; Indian Drug 6(1):17–21

Varshney V (2022) Seed extraordinaire: sabja are rich in fibers, lends to a variety of recipes. Down to Earth Magazine

Chapter 2
Aniseed, Shallot and Caraway

Abstract *Aniseed* The chapter discusses various aspects of the crop management, starting from its origin, description and distribution. It also gives importance to the production of various processed products from Aniseed. *Shallot* As usual, the chapter first discusses the origin, description and distribution of Shallot. It lays emphasis on its economic importance. The chapter further discusses its composition and other related aspects and use of Shallot. *Caraway* The chapter would discuss, in detail, the origin, description and distribution of the crop. Further discussion would target crop management, post-harvest technology and uses of processed products of caraway.

Keywords for Aniseed Description · Distribution · Crop management · Processed products
Keywords for Shallot Description · Distribution · Composition · Economic importance
Keywords for Caraway Description · Distribution · Crop management · Post-harvest technology · Processed products

Aniseed

Botanical name: *Pimpinella anisum* Linn

Syn: *Anisum vulgare* Gaertner; *Anisum officinalis* Moench.

Family: Umbelliferae

Description and Distribution

Aniseed is an annual herbaceous plant belonging to the Ajowan, Caraway, Cumin and Fennel family, Umbelliferae, a native of the East Mediterranean region. It is widely cultivated in Bulgaria, Cyprus, France, Germany, Italy, Mexico, South America, Syria, Turkey, Spain, UK and USSR. In India, it is grown to a small extent as a culinary herb or as a garden-plant. It appears that the reports of its cultivation and production are not correct since Aniseed has often been confused with Fennel (*Foeniculum vulgare*) because of the common Indian name "*Saunf*" to both. Even the commodity commonly available in the Indian market under the brand name "*Lucknow Saunf*" is called "Indian Aniseed". During 1967–68, Aniseed with INR

K. P. Nair, *A Compendium of Unique and Rare Spices*,
https://doi.org/10.1007/978-3-031-20249-0_2

Table 2.1 Trends in Indian exports and imports of Aniseed

Year	Quantity (tons)		Value (INR 000)	
	Export	Import	Export	Import
1977–78	78.70	46.00	438.00	513.00
1978–79	103.60	65.00	731.00	834.00
1979–80	172.30	125.00	1329.00	1491.00
1980–81	30.40	148.00	190.00	1555.00
1981–82	NA	81.00	NA	1525.00
1988–89	470	NA	119.50	NA
1989–90	19.10	–	160.90	–
1990–91	22.00	–	399.50	–
1991–92	32.50	–	1046.00	–
1992–93	159.20	–	1560.80	

Sources (1) Spices Export Promotion Council, Kochi, Kerala State, India. 1984. Export Statistics of Spices (1978–79 to 1980–81). (2) Spices Board, Government of India, Kochi, Kerala State, India, 1998. Spices Statistics (Book), 4th edn., pp. 1–162
NA Not Available

0.130 million was imported from France, Germany, Spain, UK, etc. It is only recently that some useful research and development work has been undertaken at the Indian Agricultural Research Institute (IARI), New Delhi, India, on the development of better varieties of Aniseed imported, which has been imported from overseas. It is grown on a small scale in Rajasthan, Punjab, Uttar Pradesh, and Odisha states, in India. It can be successfully grown as a *Rabi* (winter) crop in India.

In European countries also, Aniseed is often confused with another spice, namely Star-Anise (*Illicium verum*), which, of course, is botanically different. Aniseed bears fruits which contain volatile oil similar to that found in Star-Anise. Therefore, they have almost similar odor and flavor. Of course, Anise oil is finer in consistency and has a more delicate flavor than that of Star-Anise oil.

Aniseed is ground-gray to grayish-brown in color, 3.0–5.0 mm in length, oval in shape, and with a short stalk (pedicel) attached. Five longitudinal ridges are visible on each of the pericarp. It has a characteristic agreeable odor and a pleasant aromatic taste. Aniseed is available whole or ground. Star-Anise, though slightly cheaper than Aniseed, is commercially more important than the latter on account of the comparative price structure.

Area and Production

No reliable data are available regarding the total average of Aniseed in India. However, some available data on exports are presented in Table 2.1.

Economic Importance/Exports

The export of Aniseed (whole from India, both quantity and value-wise) during the 10 years (reported in Table 2.1), along with the corresponding import figures reveals the following:

During 1987–88, 14 tons of Aniseed were exported which fetched INR 0.78 million, with a FOB (Free On Board) value of INR 56.20/kg. Earlier during 1984–85, 21 tons of Aniseed fetched INR 0.36 million. Besides, India also imported 1670 kg of Aniseed oil worth INR 0.23 million during 1981–82. The data on country/zone-wise export of Aniseed show considerable variation both in quantum and value of export of Aniseed. For instance, during 1979–80, Aniseed worth INR 0.44, INR 0.46 and INR 0.16 million was exported to former USSR, Bangladesh and USA, in addition to export to other countries, but during 1980–81, none of these major Aniseed importing countries imported Aniseed except, the USA. Aniseed export gradually shifted from other countries to USA and Saudi Arabia. Aniseed export gradually rose to over INR 1.3 million during 1979–80, but again, declined sharply during 1980–81. Nevertheless, through systematic and sustained efforts to improve the quality, and looking for newer avenues, there is very good scope for Aniseed export from India.

Varieties

Based on the investigations conducted at the Indian Agricultural Research Institute, New Delhi, India, selection EC 22091 imported from France has got acclimatized to Indian conditions quite well. Its flowering and maturity time are uniform, and the selection is a good yielder. It contains 2% essential oil. It is suitable for cultivation in Indian plains.

Crop Management

The cultivation of Aniseed was tried at the Indian Agricultural Research Institute, New Delhi, India, farm with seed imported from Europe. After several field trials, it has been found that it can grow successfully as a *Rabi* (winter) crop. Its cultivation requirement is like those of Ajowan. Any fertile soil which can grow Ajowan will suit Aniseed, as well. The best time of planting/sowing in the northern Indian plains is mid-November of the year. Good germination can be had by sowing overnight soaked seed in well-prepared soil having optimum soil moisture. The seeds are sown in shallow furrows, followed by light planking (a plank is an Indian implement made of wood to level the surface soil).The seed should be placed at a depth of about 1 inch. After the lapse of a week, the plot should receive a light irrigation. Generally, it takes about 10–15 days for seeds to germinate. Subsequently, the crop requires to be irrigated twice or thrice only if there is a need for it, as it does not appear to stand too much soil wetness. About 35–40 cartloads of Farm Yard Manure are sufficient for a good crop. The crop matures in about 170–175 days, and a yield of 250–300 kg/acre can be obtained. Harvesting and threshing are done as in the case of Ajowan.

Table 2.2 Composition of Aniseed

Property	Content (%)
Moisture	9–13
Protein	18
Fatty oil	8–23
Essential oil	2–7
Sugars	3–5
Starch	5
N-free extract	22–28
Crude fiber	12–25
Ash	6–10
Chlorine	In traces

Post-harvest Technology

Chemical Composition/Quality

Table 2.2 summarizes the chemical composition of Aniseed.

Quality Standards

Quality standards have been laid down under the PFA Act (Prevention of Food Adulteration Act 1954) but not under Agmark (A certification mark for agricultural produce) or ISI (Indian Standards Institute). However, ISO (Indian Organization for Standardization) has prescribed an international standard; some of the important aspects are given below:

Freedom from Insects and Molds

Aniseed should be free from living insects and molds, and, should be practically free from dead insects, insect fragments and rodent contamination visible to the naked eye (corrected, if necessary, for abnormal vision) or with such magnification as may be necessary as in any particular case. If the magnification exceeds, $\times 10$, this fact should be stated in the test report.

Extraneous matter

Extraneous Matter includes

1. Inorganic: Dust, mud, stones, pieces of metal, etc.
2. Organic: All vegetable matter other than Aniseed (stems, leaves, chaff, etc.)

The different grades of Aniseed are described in Table 2.3.

Table 2.4 incorporates data on the physico-chemical characteristics of Aniseed of different grades.

Table 2.3 Different grades of Aniseed

Grades of Aniseed	Extraneous matter (% (m/m) maxm)	Proportion of shriveled, immature, damaged, broken, fruits, etc. (% (m/m) maxm)
1	1	3
2	3	4
3	4	10

Source ISO:DIS7386 (1963) Draft International (ISO) Standard for Aniseed

Table 2.4 Physico-chemical characteristics of Aniseed grown in different countries of the world

Property	Requirements		Test method
	Area 1*	Area 2**	
Moisture content (%) (m/m) maxm	12.0	12.0	ISO 939
Volatile oil content (ml/100 g) on dry basis min			
Grade 1	2.5	2.0	ISO 6571
Grade 2	2.0	1.5	ISO 6571
Grade 3	1.0	1.0	ISO 6571

Note Important areas of Aniseed production only (the list is not exhaustive, but briefly given only for information)

Area 1*: Aniseed grown here has high volatile oil contents: Algeria, Tunisia, France, Portugal and Spain

Area 2**: Aniseed grown here has low volatile oil contents: Albania, Bulgaria, Egypt, Hungary, India and Turkey

Adulteration

Aniseed is adulterated with exhausted fruits, fine earth and other small seeds and fruits. Ground Aniseed is sometimes found adulterated with ground Fennel seeds which resembles Aniseed in aroma and flavor, but considerably cheaper.

Anise oil is frequently adulterated with the lower-priced Star-Anise oil. In India, probably the oil of Fennel is sold as a substitute for Anise oil; the former can be distinguished by its lower anethole content and higher optical rotation (+11 to + 20 °C). Other adulterants used are turpentine oil, cedarwood oil and copaiba and gurjan balsam oils. Adulteration with synthetic "anethole" oil made from pine oil is also reported to be used as an adulterant.

Packaging

Aniseed should packed in clean, sound and dry containers made of a material which does not affect the product and protects it from moisture ingress and from the loss of volatile matter, for example, tin-plate containers, wooden cases or new jute bags.

Storage and Transport

Containers of Aniseed should be stored in covered premises, well-protected from the sun, rain and excessive heat.

The store room should be dry, free from objectionable odors and proofed against the entry of insects and vermin. The ventilation should be controlled so as to give good ventilation under dry conditions and to be fully closed under damp conditions. In a storage warehouse, suitable facilities should be available for fumigation.

The containers should be so handled and transported that they are protected from the rain, sun or other sources of excessive heat, from objectionable odors, and also from cross-infestation, especially, in the holds of a ship.

Processed Products

The major processed product is essential oil which is described below:

Distillation of Oil

Aniseed on steam distillation yields an essential oil, known as "Oil of Anise", which has now replaced the Anise fruits used as a pharmaceutical product and also as a flavorant. Anise oil is a colorless or pale-yellow liquid having the characteristic odor and taste of the Anise fruit.

The yield of the essential oil generally varies from 1.9 to 3.1%. Higher values up to 6% have been reported from the Syrian Aniseed. Crushing of fruits prior to distillation gives better yield of the oil. The material should be distilled soon after crushing the Aniseeds to prevent any loss of oil due to evaporation. Anise oil is a highly refractive liquid which solidifies on cooling. The congealing point depends much on the "Anethole" content and is a valuable criterion to evaluate the quality of the oil. Exposure of oil to air causes polymerization and some oxidation also takes place, with the formation of Anisaldehyde and Anisic acid.

The chief constituent of Anise oil is "anithole", which is present to the extent of 80–90%, and is mainly responsible for the characteristic flavor of the oil. The oil also contains methyl chavicol, p-methoxyphenyl acetone, and small amounts of terpenes and sulfur-containing compounds of disagreeable odor. Additionally, several minor compounds have also been identified in the Spanish oil.

Uses

Aniseed possesses a sweet aromatic taste and emits, when crushed, a characteristic agreeable odor, and, is used to flavor food, confectionery, beverages, and bakery products, "anisette" and other liquors. The fruit is considered a mild expectorant, stimulating, carminative, diuretic and diaphoretic, and is used in curing flatulence, colic, and also in the preparation of powders to cure asthma and in veterinary pharmaceuticals. Alcoholic extract of Aniseed possesses fungicidal property.

Essential oil is used in perfumery, soaps and other toilet articles and to flavor culinary preparations, confectionery beverages and liquor anisette. It is used in perfuming sachets, dental preparations and mouth washes; it is also used in the manufacture of liqueurs.

"Oil of Anise" is also reported to be used as an aromatic carminative to relieve flatulence. It is also used as an ingredient in cough lozenges in combination with liquorice. It is a mild expectorant and is used as an antiseptic and for the treatment of cholera. It may be used in the preparation of gripe water also. The "distillation water of Anise" is sold in Indian *Bazaars* (markets) as "*Araq Badian* or *Araq Saunf*" and is known for its pharmaceutical properties.

Aniseed essential oil is also used externally as an insecticide against small insects, such as, head-lice, mites and vermin. It also has fungicidal properties.

The residue left after extraction of oil from Aniseed may be used as a high grade cattle feed. It contains 17–19% protein and 16–22% fat, in addition to carbohydrates.

Ether extraction of the exhausted fruits (left after extraction of the essential oil) yields a dark green fatty oil suitable for soap-making. A hard fraction (yield up to 20%, melting point 28.5–30.1 °C) of the oil can be used as a substitute for cacao butter in the manufacture of confectioneries and pharmaceutical preparations. The fatty oil expressed from the whole fruits possesses the characteristic Anise-flavor.

Fresh leaves of the plant are used as a garnish and to flavor salads. They are eaten fresh, as "pot-herb", as they contain an essential oil and also vitamin C (8.7 mg/100 g).

Shallot

Botanical name: *Allium ascalonicum* Linn

Family: Liliaceae

Description and Distribution

Shallot belongs to the onion and garlic family and is similar in properties to those of garlic. It is a native of Palestine and is cultivated in the USA, some European countries and in Indian gardens for its bulbs and green leaves. Clusters of greenish-white or red cloves occur at the base of its hollow cylindrical leaves. A variety of Shallot, smaller and with larger number of cloves, is exported to Sri Lanka and Malaysia from Chennai, in Tamil Nadu, India.

Economic Importance

Shallot is gradually gaining importance, as is seen from the export of the same from India during 1994–95. There is much scope for its export development provided its cultivation is stepped up gradually. The encouraging export performance during 1994–95 can be seen from Table 2.5.

Crop Management

The plant is perennial which seldom produces seeds, and therefore, must be propagated vegetatively, by division of its compound bulbs, which are made up of several bulblets or cloves held together at the base. The bulbs are not encased by a sheath as in the case of garlic. The bulblets are planted in the same manner and the same season as in the case of onion sets, each developing into a compound bulb. The mature bulbs are harvested, cured and stored in the same manner as in the case of

Table 2.5 Trends in Shallot export from India

Importing countries	1992–93		1994–95	
	Quantity (kg)	Value (INR)	Quantity (kg)	Value (INR)
Sri Lanka	300,000	1,650,000	480,000	3,559,488
UAE	110,000	314,813	214,000	671,117
Singapore	45,000	236,484	19,992	159,394
Bangladesh	36,000	102,374	–	–
Malaysia	–	–	–	–

Source APEDA, Government of India, New Delhi

onions. In suitable storage, the bulbs are kept well and can be used from one season to another.

Composition and Other Related Aspects

No published information is available on Indian Shallot.

Uses

The flavor of Shallots is somewhat milder than that of onions. The chief use of Shallot is to flavor curries; both the cloves and leaves can be used for flavoring. Most of the crop is produced for sale in the green state, but some dry bulbs are also used. Shallots are also used for pickling.

Shallot is used to cure ear ache, a small piece being placed in the meatus (opening of the ear canal). It is also fried in butter and preserved in honey and used as an aphrodisiac.

In Ghana, the bulbs are ground and rubbed on the skin of children suffering from fever. This is believed to cure the fever.

Sometimes, Shallot is mixed with palm-wine and large pepper and heated in the sun, the mixture being used to cure fever. They are also used as antidotes against poisoning by snake bite and others like scorpion bites.

Important Note: The other member of the Onion family which is internationally recognized by the ISO/BIS and grown in kitchen-gardens in India is chives, *Allium schoenoprasum* Linn which is not of much economic importance and also not much information is available about it in India, and, hence, not discussed in this context.

Caraway, Black Caraway or Siah Zira

Botanical Name: *Carum carvi* Linn

Black caraway: *Carum bulbocastanum* W.Koch. *Sasali carvi*

Syn: *Apium carvi* (Linn) Crantz., *Sasali carvi* (Linn) Lam et DC.

Description and Distribution

Under the Prevention of Food Adulteration Act (PFA) of the Government of India (amended up to March 31, 1996) both Caraway (*Carum carvi* Linn) and Black caraway (Carum bulbocastanum W.Koch.) have been described in Hindi (Indian language), as "*Siah zira*" or "*Kala zir*". These two species have similar cultural requirements and are thus covered under one head.

Caraway

About 2000 years ago, Roman soldiers dispersed Caraway seed from its home in Asia Minor to most of known countries of the world. North Europeans and central Europeans, especially the British and Germans, respectively, have been using it since centuries. In India, also, it is being used since long.

Caraway or Caraway seed of commerce is the fruit of a biennial herb. It is a native of North and Central Europe and is extensively cultivated in Holland, Poland, Russia, Bulgaria, Denmark, Syria, Morocco and to a small extent in England. It has been introduced into the USA, where it is grown as a garden crop. It (Black Caraway or *Kala zira*) grows wild in India in the north Himalayan regions. It is also cultivated in the Indian plains. It is also cultivated as a cold season crop in the hills of Kashmir and Kumaon at an altitude of 2740–3660 m above mean sea level, as a summer crop.

The Caraway plant usually has a fleshy root and a slender branched stem that attains a height of 0.5–0.6 m. The compound pinnate leaves are divided into very narrow segments, the small white flowers are borne in fleet, compound umbels; the fruit, when ripe, splits into narrow, elongated carpels. A fruit is 4–6.5 mm long and curved, pointed at the ends, and with 5 longitudinal ridges on the surface. The dried fruit or seed is brown in color and has a pleasant odor, aromatic flavor, is warm with a somewhat sharp taste (because of the presence of carvone, which is a member of the family of terpenoids, naturally found in essential oils from seeds). Seeds are hard and sharp to touch. They are free from stalk ends. Caraway seed is available whole or ground. The seeds on steam distillation, yield an aromatic essential oil (yield 4–6%), which finds greater use in the manufacture of pharmaceutical products rather than as seeds, as such.

Black Caraway

Black caraway or "*Kala Zira*" (*Carum bulbocastanum* W.Koch.) has been found growing wild in the districts of Kinnaur, Lahul and Spiti, also in Tehsil (district) Pangi and Bharmour of Chamba District of India. Realizing its importance as a spice and more as a valuable seed in *Ayurvedic* (Indian system of medicine) medicines, farmers can have a gross income of INR 15,000–20,000 per hectare by growing "*Kala Zira*". It, being a perennial crop, does not require planting year after year. Through experimental trial, it has been possible to domesticate "*Kala Zira*" in higher elevations (arid zones), ranging from 1850 to 3100 m above mean sea level. The natural habitats of the plant are: low rainfall areas (150–450 mm), during the summer months and optimum temperature of 75–80 °F (24–27 °C) accompanied with heavy rainfall in winter (2–5 m). After over-wintering, the bulbs start sprouting with the onset of spring, when the snow melts. Low rainfall during the vegetative, flowering and maturity stages helps in developing flavor and quality of the seeds. The flower initiation occurs only after the bulbs attain 3 years of age. At present, only the local planting material (seed) can be multiplied. Collections from different zones in India are being made, and the germplasms are being investigated to know if there is any genetic variability. The methods of cultivation, in brief, are detailed below.

Area, Production and International Trade

The spice is known in the market by the name of the country of its origin, as for example, British, Dutch, and Mogador seed. The Dutch seed is considered to be the best among all for distilling, and, about 10% of Holland's total Caraway seed production is used to distil oil. However, no reliable statistical data about the total world production and international trade are available. The combined figures for all seed spices, however, are available. Likewise, no separate statistical data are available, yet, on the area and production of Caraway or Black Caraway in India, as well.

Crop Management

Sandy loam and well-drained soils are best for the proper development of Caraway bulb, though, it does well in loam, sandy loam, and, stony soils, as well. It can be grown in fruit-orchards in-between the rows of the plants. The fields should be plowed or dug up, as the case may be, twice or thrice to obtain a fine tilth.

Manuring/Fertilizer Application

The soil should be enriched with 200–250 quintals of well-rotted farm yard manure (FYM)/hectare before final plowing. Besides the organic manures, chemical fertilizers can also be applied, but the exact recommendation cannot be made as the trials are still underway. However, from the preliminary observations, 150 kg nitrogen, 100 kg phosphorus and 50 kg potassium have been found to enhance the yield. Urea solution at the rate of 2–3% (concentration in the solution) is sprayed at the time of bolting and seed setting, which further enhances the yield. The fertilizer requirements can even be dispensed with the application of well-rotted farm yard manure FYM and 2–3 sprays of urea solution.

Methods and Time of Sowing

The following two methods are followed to raise the crop.

By Seed Propagation

Seeds can be sown either in line or broadcast, but for the economics and germination percentage (germination percentage is low, in the latter case) line sowing is preferred. Line sowing facilitates intercultural operations like weeding. Further, it ensures uniform plant population. The seeds should be sown in lines 30 cm apart and 2.5 cm deep. About 20–25 kg of seed is sufficient to cover one hectare.

Vegetative Propagation

Caraway can also be grown by planting bulbs. The bulbs are planted 10 cm deep and 20 cm apart. The space between the lines should be 30 cm. To obtain better returns, the bulbs of 6 cm diameter are most suitable. About 170,000 bubs/ha are required.

Intercropping

The plant remains biologically active for 4 months and hence, intercropping with other crops, such as field beans, peas and buckwheat can be grown between standing rows of "*Kala Zira*", yield from which is an additional source of income to the farmer. The yield from "*Kala Zira*" can be obtained in the first week of July.

Interculture

The crop needs care in the first 3 years after planting, but only requires minimal attention once it is established well in the field. Three or four weedings are sufficient to control weeds in these 3 years, and in the subsequent years 2–3 weedings are sufficient to keep the crop in good shape, and, obtain a good harvest.

Irrigation

The crop requires, in all, 3–4 irrigations only during it entire life cycle. Of course, this depends on the type of soil it is grown in. The first irrigation is given when the plants are in the bolting stage. At the time of flowering and seed formation, water should be applied to ensure proper development of the seeds.

Harvesting and Threshing

As the crop is prone to shattering, the precise time of harvest is very crucial to minimize physical loss of seeds. It is advisable to harvest the crop when the seeds turn light brown in color for commercial purposes and dark brown in color when it is used for sowing purposes. The crop may be harvested at night or in the morning, when the dew is still on. Threshing is physically done using bamboo sticks.

Yield

Average yield of Caraway is 8–10 quintals/ha (100 kg = 1 quintal).

Plant Protection: Insect Pests

Controlling Aphids: Although no major insect pest affects the Caraway plant, Aphids attack the umbels and suck the sap of flowers which leads them to dry up. Spraying 0.02% Folidol controls Aphid attack.

Cut Worms: The emerging succulent bolts of the Caraway plant are attracted by Cut worms. Dusting with Aldrin dust at the rate of 50–60 kg/ha proves effective in controlling Cut worm.

White Grubs: The bulbs are damaged by White Grub. The White Grubs can be destroyed either by exposure to the sun or by application of Aldrin (50%) dust at the rate of 50 kg/ha. It should be applied at the time of preparatory tillage.

Plant Pathogens: Zira Blight: The blight damages the crop heavily. Spraying with Dithane Z-78 (0.2% concentration) spray will control the pathogen. Spraying should commence before the crop begins to bolt.

Table 2.6 Composition of Caraway seeds

Property	Content
Moisture	4.5%
Protein	7.6%
Fat	8.8%
Fiber	25.2%
Carbohydrates	50.2%
Ash	3.7%
Calcium	1.0%
Phosphorus	0.11%
Potassium	1.9%
Sodium	0.02%
Vitamins (mg/100 g)	
Vitamin B1 (thiamine)	0.38%
Vitamin B2 (riboflavin)	0.38%
Niacin	8.1%
Vitamin C (ascorbic acid)	12.0%
Vitamin A	580 IU (International Units)
Calorific value (food energy)	465 cal/100 g of the spice

Harvesting

In India, the caraway fruits are collected before ripening. Well-ripened fruits may also be harvested in the very early morning, when the plants are bathed in the morning dew; if harvested when it is hot and the sun is up, many seeds would fall even before harvest. The plants are dried, the fruits are threshed out and dried, cleaned and stored in gunny bags. The yield is very variable. It may be as high as 2500 kg/ha on rich soils, but the average yield reported from Holland varies from 750–2000 kg/ha.

Post-harvest Technology

Packaging and Storage

The procedure is the same as in the case of Ajowan or Aniseed (described earlier in this book). The seeds are packed in small bags. Caraway is mostly sold whole. It has good storage quality as in the case of Black pepper.

Composition

The composition of Caraway seeds and fruits is given in Tables 2.6 and 2.7.

Adulteration

The commercial Caraway seed is often adulterated with spent or exhausted (after extraction of the essential oil) Caraway seeds, Cumin seeds, Carum stems, Gravel

Table 2.7 Composition of Caraway fruits

Property	Content (%)
Moisture	11.5–15.5
Ash	5.5–6.7
Water-soluble resin	6.2–10.1
Crude fiber	17.5–22.33
Nitrogen	5.9–6.4

Table 2.8 International (ISO) physico-chemical requirements of Caraway seed (whole)

Physico-chemical characteristics	ISO requirements
Extraneous matter (%) (maximum)	1.0
Moisture content (%) (m/m) (maximum)	13.0
Total Ash % (m/m) on dry basis (maximum)	8.0
Acid-insoluble ash %(m/m) on dry basis (maximum)	1.5
Volatile oil content ml/100 g on dry basis (minimum)	2.5

Source ISO 5561-1981. International (ISO) Standard for Caraway (whole)

and Earth (soil). Caraway powder is adulterated with cheap spice seeds such of similar flavor.

Quality Standards (PFA/BIS/ISO—read explanations earlier in this book for expansion of abbreviations):

The quality standards prescribed under PFA Act for Caraway and Black caraway are almost identical in all physico-chemical parameters. The BIS and ISO have also been formulated as per national and international standards for caraway seeds (whole and ground). The ISO specifications are summarized in Table 2.8.

Processed Products

The two important processed products of Caraway whole seed are (1) volatile oil and (2) carvone, isolated from oil of caraway. They are briefly discussed below:

Volatile Oil

Caraway oil is steam-distilled from fresh whole seeds. It is a colorless or pale-yellow oil with the following physico-chemical characteristics which are given in Table 2.9.

Caraway grown in Kashmir (India) gives an oil, conforming to the British Pharmacopoeia standard. The volatile oil contains a mixture of ketone, carvone (specific gravity 0.850), a terpene, formerly called carvone, but now recognized to be dl-limonene and traces of carvacrol. Pure carvone is prepared by decomposing crystalline compounds of carvone with hydrogen sulfide.

Table 2.9 Physico-chemical characteristics of seed (whole) Caraway

Property	Content
Specific gravity at 15 °C	0.907–920
Optical rotation at 25 °C	70°–85°
Refractive index at 25 °C	1.484–1.498
Solubility in 2–10 volumes 80% alcohol Carvone content	45–65%

Adulteration in Caraway Oil

Commercial oil is often adulterated with decarbonized oil. The official standards require that the oil should contain not less than 53% and not more than 63% carvone.

Decarbonized Oil

It consists of limonene with traces of carvone and is sold in the market as light oil of Caraway. It finds use in scenting cheap soaps. The characteristics of light oil are: specific gravity: 0.848, optical rotation: 130°.

Caraway Chaff Oil

As inferior oil, Caraway chaff oil is said to be distilled from the husks and stalks which remain after threshing. The dried, exhausted and pulverized Caraway chaff contains 20–23.5% crude protein (of which 75–85% is digestible) and 14–16% fat. It can be used as a cattle feed.

Uses

As a Food Flavorant

Caraway is widely used as a spice for culinary purposes and to flavor rye-bread, biscuits, cakes and cheese. It is also used in the manufacture of "*Kummel*" cordial and as an ingredient to season sausages and pickling spice, soups and meat stews.

Medicinal Uses

Caraway is a mild stomachic and carminative, occasionally used in flatulent colic and as an adjuvant or corrective for medicines. However, its volatile oil is employed more often than the fruit itself.

Carvone isolated from Caraway oil is used as anthelmintic in hookworm infection in humans. Caraway oil is chiefly used to flavor food and in medicine as a carminative. It is also used to correct the nauseating and groping effects of medicines. For the treatment of scabies, a solution containing 5 parts each of alcohol and oil of Caraway in 75 parts of castor oil is recommended. Caraway seed oil is used in oral preparations for overcoming unpleasant odor or taste.

It is also used for scenting cheap soaps; when used as a scenting agent, the soap gives the characteristic full note.

Chapter 3
European Dill and Indian Dill (Sowa)

Abstract The chapter would first discuss the description and distribution of the crop worldwide. It would lay emphasis on the anatomical parts of the essential oil Dill including essential oil.

Keywords Description · Distribution · Processed products · Essential oil

Botanical name: *Anethum sowa* Roxb.ex.Flem.

Family: Umbellifereae

Description and Distribution

The genus *Anethum* comprises three species, two of which are recognized for commercial use and yield Dill oil used in medicine. These two species are:

1. *Anethum graveolens* (European Dill)
2. *Anethum sowa* (Indian Dill).

European Dill is indigenous to Europe and is cultivated in England, Germany, Romania, Turkey, USA and former Soviet Union. Until recently, it was not grown in India. However, efforts have been made to acclimatize it in Jammu (Kashmir, India) according to the Regional Research Laboratory (RRL, Jammu) and the Indian Agricultural Research Institute, New Delhi, India, Bhubaneshwar (Odisha State), and its exports also have started since 1988–89. The trends in exports are given in Table 3.1.

The Indian Dill (*Anethum sowa* Roxb.ex.Flem.) is often regarded as a variety of *Anethum graveolens*. However, the ripe dried seeds of Indian Dill are larger, bold, good looking, but less broad than those of European Dill. Their dorsal ridges are paler in color. Color is almost similar, but again, they differ in respect of composition of their essential oil content.

Botanical Description

A. sowa is an herbaceous annual, with pinnately divided leaves, found practically throughout India. It is also cultivated as a cold-weather crop in many parts of India. The ripe, light brown seeds emit an aromatic odor faintly resembling that of Caraway and possess warm aromatic and slightly sharp taste, almost akin to Caraway seeds.

K. P. Nair, *A Compendium of Unique and Rare Spices*,
https://doi.org/10.1007/978-3-031-20249-0_3

Table 3.1 Trends in annual export of Dill from India

Year	Quantity (tons)	Total value (INR 000)	FOB unit value (INR/kg)
1988–89	724	7912	10.93
1989–90	739	7523	10.18
1990–91	474	5086	10.73
1991–92	634	9734	15.35
1992–93	549	10,150	18.49

Source Spices Board, Government of India (1994, pp. 1–162)
FOB: Free On Board (shipping term)

The following are the anatomical parts of the essential oils of Dill.

Dill Herb: Fresh "weed" (stalks, leaves and seeds).

Dried Dill Herb: Dried aerial parts of the plant.

Dill Seed-Dry Fruit.

Dill Herb/Weed OiI: Obtained by steam distillation of fresh whole plant (stalks, leaves and seeds) or dried leaves and seeds (matured fruits).

Dried Seed Oil: Derived from steam distillation of crushed dry, mature/ripe fruits.

Dill is of value for both its leaves and seeds. Dill leaves are also known as "Dill Weed", fresh or dried. When fresh, the leaves have a pleasant aromatic color and warm taste. Dill seed with its pleasantly aromatic and warming flavor, laced by woody and menthol notes, is excellent for flavoring and seasoning of several food items.

Historical Aspects

We actually do not know how old Dill is, but early Babylonian and Assyrian herbals listed it, and it is known that the Romans gave their gladiators a tonic of Dill in the belief that it was a stimulant. Much of Dill's ancient history is based on the assumption that the term "Aneth" was earlier referred to Dill. It is under this pseudonym that it appears in the Bible—in Mathew, Chapter 23.

By the Middle ages, Dill was widely used in cooking, but an interesting anomaly was beginning to develop. This plant, which was native to the Mediterranean region, was becoming a staple food additive to North European cooking, but forgotten in the Southern European regions. French and Italian cook books hardly mentioned it, whereas German, Scandinavian and Russian dishes made it an essential ingredient in various food preparation.

The most famous culinary use of Dill was as the "Dill Pickle", at least 400 years old. Parkinson in his Paradisi in Sole (1629) noted "It (Dill) is also put among pickled cucumbers where it doth very well agree, giving to the cold fruit a pretty spicie taste or relish". Evelyn (1629–1706) also praised "gherkins muriated with seeds of dill".

And Addison (1672–1719) wrote: "I am always pleased with that particular time of the year which is proper for the pickling of dill in cucumbers".

The Western World, Russia and parts of Middle East, grow the Dill classified as *Anethum graveolens*—the so-called true Dill. This is believed to be a native of the Mediterranean region and Southern Russia. India and certain regions in Asia grow a type of Dill called *Anethum sowa* or the Indian Dill, which is native to Northern India. The principal difference is that the Sowa seeds are especially bold (large and good looking). They also tend to be longer and narrower in shape, and their oil has a somewhat different profile.

Economic Importance and Exports

Though Dill was acclimatized in India recently, as a result of intensive research and development efforts by the concerned authorities, named above, especially RRL (Regional Research Laboratory, Jammu and Kashmir State, India), CSIR (Council of Scientific and Industrial Research, a Government of India undertaking in New Delhi, India) and ICAR (Indian Council of Agricultural Research, New Delhi, India) its exports started in 1988–89. About 549 tons of Dill seeds were exported from India earning the country INR 10.15 million at an FOB (Free On Board) price of INR 18.49/kg (refer Table 3.1). Though tonnage exported by 1992–93 decreased by 24.17% (from 724 tons in 1988–89 to 549 by 1992–93), the total value of the produce increased by 28.24%, primarily because of the enhancement of FOB by 69.17%, from INR 10.93 to INR 18.49. This demonstrates the remarkable potential for Dill export from India.

Area and Production

No reliable statistics are available on the above aspects of Dill seed, yet, as reported by the Department of Economics and Statistics, Ministry of Agriculture, Government of India. India produced about 4963 tons of *Anethum sowa* during 1993–94. Thus, there seems to be excellent scope to improve further the processing/manufacture of Dill Oil and Carvone, of which, it is quite a good source. Caravone is used as a synergist and as an insecticide, apart from its use as a flavorant and perfumery functionality.

Crop Management

The available scanty information on crop management of Dill is summarized below.

Soil: Dill thrives best on sandy loam soil. Its cultivation is similar to those others of Umbellifereae family.

Sowing

The optimum time of sowing varies with different climatic zones. In India, the seed is sown from September to October in most parts. Sowing is done in well-prepared land, which is incorporated with FYM (Farm Yard Manure) at the rate of 20–30 cartloads/hectare. The seeds are sown 30 cm apart in rows spaced at 45 cm.

Table 3.2 Physico-chemical composition of Dill

Property	Content
Moisture	4.5–7.5% (average 6.2%)
Volatile oil	1.5–4.0% (average 2.7%)
Mineral matter (Total ash)	5.89–10.54% (average 9.09%)
Acid-insoluble ash	0.55–2.71 (average 1.20%)
Extraneous matter	3.16–12.93% (average 4.93%)
Damaged seeds	0.0–3.05% (average 0.71%)
Cut seeds	0.03–0.39% (average 0.13%)
Immature and shriveled seeds	3.32–15.78% (average 7.23%)

Seed Rate: Seed rate is 8–10 kg ha. The seedbeds, after sowing, are covered with a thin layer of soil, and planking is done to level the field.

Interculture: Seedling emergence is observed, generally, about 8–10 days after planting. When the seedlings attain a height of 10 cm, they are manually thinned, retaining only the healthier plants. One or two manual weedings are done, and 2–3 irrigations are given.

Harvest: The crop is harvested when the first flower heads turn brown, are dried in the shade and threshed by hand or with a small wooden stick to disengage the seeds from the stem, like in any other case of small spices, like cumin.

Yield

Recent investigations at Punjab Agricultural University, Ludhiana, Punjab State, India, on the growth, yield and nutrient uptake of Dill (*Anethum graveolens* Linn.) have shown that the maximum Dill seed yield could be obtained when the crop is sown on or about 15th October in lines 30 cm apart, using 10 kg seed/ha and supplemented with 90 kg nitrogen/ha (Avtar Singh and Randhawa, 1987). The same authors in 1995 have published an excellent review on the crop management technology of Dill (Singh et al. 1992) and have concluded that no information is, yet, available in India on the effectiveness of any insecticides/fungicides or weedicides in controlling the pest (both fungal/insect) or weeds, respectively.

Post-Harvest Technology

Quality Grading/Standards

No Agmark or PFA standards have yet been laid down for Dill seed.

Composition. Table 3.2 describes the composition of Indian Dill seed (*Anethum sowa*) averaged from 10 representative samples of commercial significance.

US import of Dill seed has been steadily increasing since 1992–93 from 722 tons in 1992 to 830 tons in 1993 and further to 947 tons in 1994 as can be seen from Table 3.3. India is the leading supplier of Dill seed to USA with 68% in 1994, followed by Egypt with 7% and Pakistan with 5%.

Table 3.4 details the composition of *Anethum graveolens*.

Table 3.3 A comparative picture of Dill import into USA from different countries

Source of export (country-wise)	1992		1993		1994	
	Quantity	% Share	Quantity	% Share	Quantity	%Share
Egypt	66	9	83	10	70	7
India	540	75	565	68	643	68
Pakistan	53	7	110	13	47	5
Others	63	9	72	9	187	20
Total US imports	722	100	830	100	947	100

Notes Quantity in tons

Table 3.4 The physico-chemical composition of *Anethum graveolens* (European Dill)

Property	Content (%)
Moisture	6.6
Protein	13.1
Fat	17.9
Crude fiber	20.7
Carbohydrates	35.7
Mineral matter (total ash)	6.0
Calcium	1.6
Phosphorus	0.21
Potassium	1.1
Iron	0.012
Sodium	0.01
Vitamins (mg/100 g)	
Vitamin B1	0.42
Vitamin B2	0.28
Niacin	2.8%
Vitamin C (ascorbic acid)	12.0
Vitamin A	175 (international units/100 g)
Calorific value (food energy)	435 cal/100 g

Packaging

The seeds after drying and cleaning are packed in gunny bags and kept in a cool and dry place.

Buying and Using Dill Seed/Weed

The pickle industry is the largest user of Dill, and, for this, Dill-weed (referring to Dill leaves) oil is the main choice. Dill-weed oil can be treated to disperse easily. This product flavors pickles evenly and does not discolor the liquid. When available, Dill seed oil and Dill oleoresin are also used in pickling. Whole and ground Dill seeds are used commercially in a wide variety of products, including sausages, baked goods, condiments, cheese, curry powder and dry mixes for sauces and dressings. Dill weed has become an important herb for the spice shelf, but it is also used in commercial applications similar to the seed. Carvone content is the key to Dill buying, especially in the extractives domain.

Handling and Storage of Dill

Dill and its extractives should be stored under cool, dark, dry conditions to protect against loss of flavor and fragrance. As with any highly aromatic herb, Dill weed should be further protected against flavor loss by making certain that the container is always air-tight, tightly closed, after every use. Under good storage, Dill has shelf-life of 6–9 months.

Processed Products

There are six types of processed products made out of Dill. They are:

1. Dill seed
2. Dill seed powder
3. Dill leaves (normally called Dill weed)
4. Dill seed oil
5. Dill weed oil
6. Dill oleoresin.

In India, only Dill seed powder and Dill seed oil are traded. They are briefly described as follows:

Dill seed

Most of Dill destined for export to USA comes from India, as given in Table 3.5. The seeds of *Anethum sowa* have a pleasant Dill flavor, but somewhat stronger than that of Dill seed. The seeds are elliptical, slightly humped and with clearly defined ridges running length-wise. The tan-colored Dill seed's most distinguishing visual point, however is a thin, yellowish edge, which frames it all around. When harvested, the seeds have a hair-like strand, or whisker, at one end. In years past, there was a "dewhiskered" grade designation in the spice trade, but today Dill is so routinely "barbered" at the source that the term is no longer needed. Dill seed yield between 2 and 4% of essential oil. Please refer Table 3.5.

Dill Weed (Dill Leaves)

None seems to know why the Dill leaves are known as "Dill Weed". But it is a term widely used in the world spice market. Growers speculate that it is because Dill is such a hardy plant, that its seeds itself and thrives about anywhere it finds a well-drained soil with plenty of sun around. At one time, the "weed" designation was literal because most of the leaves came from the wild plants, but cultivation is now more common. The USA, primarily, California, is the major producer of Dill weed for the spice shelf. Egypt is another important source of Dill supply. When the objective is for the Dill weed for use as a spice rather than for extraction, the top

Table 3.5 Trends in export of Dill seed, Dill powder and Dill oil from Indian Dill

Dill seeds and Dill products	1990–91		1991–92		1992–93	
	Quantity (tons)	Value (INR 000)	Quantity	Value (INR 000)	Quantity (tons)	Value (INR 000)
Dill seed	445.5	4788.7	603.0	9315.9	544.2	10,069.5
Dill seed powder	28.1	297.7	30.8	418.5	4.5	80.0
Dill seed oil	–	–	Negative	23.2	Negative	33.2
Total exports	473.6	5086.4	633.8	9757.6	548.7	10,182.7

Source Spices Board, Government of India (1994, pp. 1–162)

20 cm of the plant is cut/harvested and the stems are removed. Some growers feel that it should be harvested before the plants flower. Others opine that it is preferable that the leaves are harvested later. In general, the leaves have a mellower, fresher flavor than the seeds. Dill weed has become increasingly popular in the USA in recent years. Indian growers/traders may also harness the immense potential of exporting Dill weed.

Dill Extractives

Dill-Weed Oil: This is the major Dill extractive. Most of it is produced in Washington, Oregon, Idaho and Western Canada. Some also comes from Texas. It is a field-distilled product, since it is felt that the best quality comes from freshly harvested leaves of Dill. For this purpose, Dill has been commercially grown in the USA for more than 30 years. As compared to Dill-Seed Oil, the Dill-Weed Oil has a sharper, cleaner, fresh Dill flavor and one does not seem as "oily". The Dill oils are bought on the Carvone-content basis, but the Dill-Weed oil normally has less Carvone content than the Dill-Seed oil.

Dill-Seed Oil: Food Chemical Codex lists the following three types of spice oils:

1. Dill-Weed Oil
2. American-types (Dill-Oil, Dill-Herb Oil, and American type)
3. Dill Seed-Oil, European-type, Dill-seed oil and Indian type.

Very little Dill-seed oil is being used in the USA. Small amounts are imported from Europe and India (though spice buyers in India feel that most of the product that comes through Europe today is actually Indian types). The Dill seed oil has higher Carvone content which gives it a stronger, more strident flavor. During 1992–93, India exported Dill-seed oil worth over INR 33,200, while during 1991–92 the corresponding earnings were about INR 23200—a remarkable increase of 43.10% in just one year. This has clearly demonstrated the huge potential in Dill oil sale from India.

Table 3.6 Comparative physico-chemical characteristics of essential oils from Indian and European Dill

Important physico-chemical characteristics of essential oils	Sowa (Indian dill) oil	Dill (European oil)
Specific gravity at 20 °C	0.9675	0.896
Refractive index at 25 °C	1.4943	1.473
Optical rotation at 25 °C	47.6	70
Carvone content (%)	19.5	45.9

Dill Oleoresin

This is prepared from the Dill seed unlike Dill weed and seed oils, which contain only the volatile fractions. The Oleoresin fraction contains both volatile and nonvolatile extractives. It is now becoming gradually popular in the world trade. Indian Oleoresin processing industry may consider its manufacture for export which is totally absent as at present.

Essential Oil

On steam distillation, Indian Dill yields 1.5–4.0% volatile oil, while European Dill grown under Indian conditions yields up to 2.5–4.0% volatile oil. At the Indian Agricultural Research Institute, New Delhi, India, a new selection has been evolved and released for commercial cultivation in India. It yields as high as 6.38% essential oil. Such new selections should be further encouraged for cultivation. This is primarily important because India still imports about INR 20,000 worth of Dill from overseas. It should not be difficult to be completely self-sufficient for India in Dill oil by increasing the area under this new selection. The comparative physico-chemical characteristics of the two essential oils are given in Table 3.6.

From the data contained in the above table, it is very clear that both the specific gravity and refractive index are higher in Indian (Sowa) Dill compared to the European Dill. The significant difference is in the caravone content, which is more than a hundred-fold in the European Dill compared to the Indian Dill. The higher specific gravity and refractive index in the Indian Dill is caused by, most likely, the presence of a larger quantity of Dill-apiole in Sowa oil. After removal of the apiole, it approximates in respect of specific gravity and optical rotation, with that of European Dill, but not in the case of Carvone content. Part of the Sowa-oil is heavier than water and sinks in the receiver during distillation. This heavier fraction, with a boiling point of 230–285 °C, consists chiefly of Dill-apiole which is isomeric with Parsley apiole. A recent useful investigation done in England on the detailed examination of 16 samples of essential oils from Dill seeds obtained from England, Germany, India, Romania, Turkey, USA and former USSR, using chromatography has revealed that Dill-apiole is absent in European-Dill, even when grown under Indian conditions. And Dill-apiole is present in Sowa oil even when grown under European conditions.

The method of harvesting and drying significantly affects yield and quality of the oil. The work done at the Indian Agricultural research Institute in New Delhi, India, has shown that the yield of oil was 2.88% when seeds were dried in shade, but only 1.03% when dried in the sun. Harvesting at the just ripe stage had higher oil content than over mature seeds had.

The Sowa herb yields 0.06% of essential oil, which has a higher proportion of terpenes (aphellandrene), but no Carvone. The European and American Dill herb oils contain both Carvone and alpha (please use symbol)-phellandrene, although the Carvone content (about 20%) is much lower than that of seed oil.

Recently, it has been shown that the oil content of Dill is not affected by sowing time, but the maximum yield was obtained when the seeds were sown in September (Randhawa and Singh 1991).

Uses

In Pharmaceuticals: The seeds are well known for their medicinal properties, mostly due to the essential oil in them. Both seeds and oil enter into the composition of various indigenous medicinal preparations. The essential oil Dill oil, or its emulsion in water, "Dill Water" is considered to be an aromatic carminative, specially useful in controlling human flatulence, colic and hiccups in infants and children.

As A Food Flavorant: Dill Weeds are used, both whole and ground, as a condiment in soups, salads, processed meats, sausages, spicy table sauces, sauerkraut (a common German preparation), and, in particular, in pickling. Dill stems and blossom heads are used for Dill pickles and to flavor soups. Ground seed is an ingredient in seasoning. Sometimes, it is sued as a substitute for Caraway.

The green herb is used as a pot-herb and as a flavoring agent. The essential oil is also used in the manufacture of soups.

Dill-Waste (Cake): The dried residue left after distillation of the essential oil from the Dill seeds of *Anethum graveolens* contains 16.8% fat and 15.1% protein. It may be used as a cattle feed, after recovering the fixed oil

As Synergist for Insecticides: The Indian Agricultural Research Institute, New Delhi, India, has developed two patents which employ some of the constituents of Sowa, with minor modifications, for use as synergists for insecticides.

References

Randhawa GS, Singh A (1991) Effect of sowing time and harvesting stage on oil content, herbage and oil yield of dill (*A. graveolens*). Ind Perfumer 35(4):204–218

Singh A, Randhawa GS (1987) Growth, yield and nutrient uptake of dill (*Anethum gaveolens*) under same agronomic practices (date of sowing, seed rates, spacings and plant nutrition). In: Pruthi JS (ed) National symposium on spice industries, 9–10 Apr 1997. AFST(1) New Delhi, India, Abstract. No. 1.31, p 26

Singh A, Mahey RK, Randhawa GS (1992) A review on production technology of dill (*Anethum graveolens* Linn.). Ind Perfumer 36(1):45–53

Spices Board, Government of India (1994) Spices statistics (updated) 2nd edn., pp 1–162

Chapter 4
Poppy Seed

Abstract The chapter would at first discuss the description and distribution of Poppy seed worldwide. Since the crop is a source for opium manufacture, a discussion at length would be on its sociological consequences.

Keywords Description · Distribution · Post-harvest technology · Opium manufacture

Botanical Name: *Papaver somniferum* Linn.

Family: Papaveraceae.

Description and Distribution

Poppy seed (*Papaver somniferum* Linn.) is just what one might think—the mature seed of a beautiful flower, known and used for at least 5000 years. Native of the Mediterranean region, as are so many of spice seeds and herbs, Poppies were taken to India and the East by Islamic missionaries, more than a thousand years ago.

Poppy seed is so tiny that one pod may contain up to 30,000 seeds and it takes some 9,000,000 seeds to weigh one pound! Each seed has a tiny droplet of nut-sweet oil which explains its flavor appeal. Although Poppy seed comes from the same plant which produces opium, the seed cannot form until after the plant has matured to the point where it has lost all of its opium potential. Therefore, there is absolutely no narcotic content in Poppy seed.

Poppy seed is grown in the Netherlands, Poland, Denmark, Sweden, the Balkan countries, Turkey, Argentina and, to some extent, India.

Poppy is cultivated either for the manufacture of opium or for its seeds. In India, variety *album* (*Papaver somniferum* Linn. var. *album*) of Poppy with white seeds has been cultivated for many years for the production of seeds under license in Dehradun and Tehri Garhwal districts of former Uttar Pradesh (now Uttarakhand), in India. It is also grown in Jalandhar, Patiala, Kapurthala and Hoshiarpur of Punjab State, in India, and also in some districts of Madhya Pradesh, in India. The continuation of cultivation of Poppy for Poppy heads and Poppy seeds was reviewed by the All India Narcotics Conference in 1956, and, on its recommendation, the State Government of Punjab banned the cultivation of Poppy for Poppy heads for the manufacture of

K. P. Nair, *A Compendium of Unique and Rare Spices*,
https://doi.org/10.1007/978-3-031-20249-0_4

Table 4.1 Trend in the annual export of Poppy seed from India

Year	Quantity exported (tons)	Value (INR 000)	FOB unit value (INR/kg)
1988–89	69	1366	19.80
1989–90	64	2.015	31.48
1990–91	9	714	79.33
1991–92	16	1369	85.56
1992–93	51	2019	39.59

Source (1) DGCI & S. Kolkata, Government of India; (2) Spices Board, Kochi, Kerala State, India
Note FOB: Free On Board

opium for a maximum period of 4 years, by which time, it was expected that necessary developmental measures would be taken to enable the cultivation of alternative crops *in lieu* of opium Poppy. In India, best seeds are obtained when the capsules have not been incised for extraction of opium. The crop is reported to yield 220-275 kg/ha of seeds (200–250 £/ac). In Europe, variety *nigrum* (*Papaver somniferum* Linn. var *nigrum*) which has slate-to-blue colored seeds, and known as "Maw Seeds", is exclusively cultivated for this purpose.

Area, Production and Exports

No official estimates of area, production and exports of Poppy seed are available. However, it is learnt that opium Poppy plays a significant role in the economy of cultivators in Madhya Pradesh, in India. It is being cultivated in Mandsaur, Ratlam, Rajgarh and Dhar districts of Malwa in nearly 40,000–45,000 ac. The crop cultivation is concentrated in Mandsaur district in about 25,000–30,000 ac, scattered all over the district wherever irrigation facilities are available, as shown in Table 4.1.

US Imports of Poppy Seed

USA is an important market for Poppy seed which has been growing from 4887 tons in 1992 to 5162 tons in 1993 and 5168 tons in 1994, which works out to an increase of 5.75% in tonnage exported in 2 years. In fact, the FOB value (INR/kg exported) works out to 338.77% between 1988 and 89 when India started exporting Poppy, which peaked in 1991–92 (Table 4.1). Australia and Netherlands are the main suppliers. Turkey supplied 355 tons in 1994 while India's share was a meager 3 tons.

Crop Management

Opium/Poppy is a short-duration crop which brings in high monetary returns in four to four-and-a half months. The cultivation of this crop requires a good knowledge of all agricultural operations as well as the soil management and climate requirement.

Seed Preparation

The most suitable soil for this crop is the well-drained highly fertile, light black cotton soil, having good percentage of fine sand. The field is plowed when it is quite dry following the rains so that big soil clods are pulverized. The clods are broken and the soil is well pulverized with repeated plowings, followed by soil leveling. If the field is plowed wet, the soil becomes somewhat cloddy, and friable texture is not obtained. In all, 6 or 7 plowings are needed. Before sowing, the field is divided into small beds, 1.8 m long 0.91 m wide by raising small bund 3–6 in. high, for giving controlled irrigations.

Sowing

Sowing is done in the month of November by broadcast of seeds, 1.5–2.0 kg/ac. The seed is mixed in the soil with the help of a rake. One light irrigation is given just after sowing. The seed starts to germinate in a week's time. Generally, the cultivators keep their own seed from the previous year's crop, for sowing purpose.

Irrigation

Water is an important factor to get good yields. The crop requires well-planned, judicious irrigations during the growth period. Faulty irrigations may ruin the crop. The first irrigation is given just after sowing and a second irrigation is given 6 days later, when the seeds start to geminate. The third irrigation is given 12 days after completion of full germination, followed by another 15 days later. Thereafter, the crop is irrigated at every 8–10 days intervals, and the crop requires, in all, 10–12 irrigations during its total life cycle.

Inter-culture and Weeding:

Inter-culture, weeding and thinning are important operations to keep the Poppy crop in good stead. Thinning is of much importance, as the seeds are small and broadcast. Healthy plants are set at a distance of 6–9 in. Weak and diseased plants should be culled out. The crop should be kept absolutely weed-free. Every irrigation is followed by hoeing (with a native Indian implement with a wooden handle and a cutting blade called "*khurpi*", in the Indian—Hindi—language).

Manuring

No standard practice is followed in manuring the Poppy crop. Cultivators apply a basal dose of 25 cart loads of FYM (Farm Yard Manure)/acre during September and October, and plow it in. The FYM is transported before the onset of rains and stored in the field. Tank silt is also applied wherever it is available. Very little inorganic fertilizers are used. Some cultivators apply Ammonium sulfate at the rate of 100 kg/ac and superphosphate at the rate of 50 kg/ac at the time of sowing. Fuel ash is also applied to the crop. A top dressing of Ammonium sulfate is applied when the crop is a month old.

Rotations

The maize-poppy, chilly (red)-poppy or sunhemp-poppy rotations are generally followed by farmers. In rare cases, the field is kept fallow in the *kharif* (summer season) season. It is a common practice not to irrigate the preceding crop as this is considered harmful to the Poppy crop.

Extraction of Opium

The crop flowers in January and February and the fruits are fully ripe by February–March, when incisions are made on the fruits with the help of lancets during afternoons. The milky sap oozes out of the fruits and solidifies, and is removed the next day in the morning. The operation is done 3–4 times on each capsule at regular intervals. The incisions are made in the field nearly 6–8 times, depending on the maturity of the capsules. The yield per acre of opium varies from 8 to 10 kg (according to information supplied by the cultivators). However, no reliable data have been collected or reported on yield.

Growing of garlic on the bunds of the Poppy field was found to be profitable as garlic did not adversely affect the latex yield from opium.

Yield of Seeds

Nearly 223.94 to 373.24 kg of seed/ac are produced. As the production and area are controlled by the Central Excise, Government of India, opium is sold to the Excise Department and the price is fixed by this agency. Evolving improved varieties and suitable agronomical practices can give higher yields of opium of better quality and bring in more money to the opium farmers and earn the country valuable foreign exchange (hard currency) to the nation concerned.

Post-harvest Technology

Packaging and Storage

The seeds are packed in ordinary gunny bags or on a smaller scale in friction-top tins.

Quality/Composition of seeds

Table 4.2 summarizes the chemical characteristics of the seeds.

The white seeds are small; 1000 seeds weigh from 0.125 to 0.5 g. Table 4.2 summarizes the analysis of 5 types of Poppy seeds.

The seeds have a high protein content, the major component being a globulin, which accounts for 55% of the total nitrogen. The amino acid makeup of the globulin is similar to that of the whole seed protein and is as follows, given in Table 4.3.

The proteins are deficient in lysine and methionine. At 10% level of intake, they have a biological value of 57.9% and a digestibility coefficient of 81%.

Quality Standards

Quality standards for Poppy seed (*khas khas*) have been laid down under the Agmark, ISI and PFA Act 1954, amended and up-to-dated.

Processed Products

Poppy Seed Oil

Poppy seeds contain up to 50% of an edible oil, which is extracted by either cold- or hot-pressing procedure. The oil is odorless and possesses a pleasant almond-like

Table 4.2 Quality/composition of representative Poppy seeds

Property	Content
Moisture	4.3–5.2%
Protein	22.3–24.4%
Ether extract	46.5–49.1%
N-free extract	11.7–14%
Crude fiber	4.8–5.8
Ash	5.6–6.0%
Calcium	1.03–1.45%
Phosphorus	0.79–0.89
Iron	8.5–11.1 mg/100 g
Thiamine	740–1181
Riboflavin	756–1203
Nicotinic acid	800–1200 μg/100 g
Carotene	Absent
Iodine	6 μg/kg
Manganese	29 mg/kg
Copper	22.9 mg/kg
Magnesium	15.6 g/kg
Sodium	0.3 g/kg
Potassium	5.25 g/kg
Zinc	130 mg/kg
Lecithin	2.80%
Oxalic acid	1.62%
Pentosans	3.0–3.6%
Narcotine and an amorphous alkaloid	Traces
Enzymes-Diastase, Emulsin, Lipase and Nuclease	Traces

Table 4.3 The amino acid makeup of the globulin in Poppy seed

Property	Content
Arginine	10.4 (g/16 gN)
Histidine	2.9
Lysine	1.5
Tyrosine	4.7
Tryptophan	2.0
Phenylalanine	4.1%
Cystine	2.0
Methionine	2.3
Threonine	4.2
Valine	7.1

Table 4.4 Ranges of constants of Poppy seed oil

Property	Range
Specific gravity at 15/20 °C	0.924–0.927
Refractive index	1.467–1.470
Iodine value	132–142
Saponification value	188–196
Acid value	3.13%

The oil also contains vitamin F which helps in curing problems of human skin

taste. In India, the oil is generally extracted by cold-pressing the seed in small presses in homes or small establishments (oil yield about 20%). Raw cold-pressed oil is pale to golden yellow in color. Poppy seed oil has the following ranges of constants (Table 4.4).

Uses

Poppy Seeds (*khas khas*)

The Poppy seeds are utilized as food and as a source of fatty oil. They are considered nutritive and are used in breads, cakes, cookies, pastries, curries, sweets and confectionery. Seeds are demulcent and are used in the form of emulsion, as an emollient and specifically against obstinate constipation and in catarrh of the bladder. The white seeds are sometimes used in pharmaceuticals. Poppy seeds or seed meal also find use in the production of lecithin (yield 0.67–0.91%).

Poppy Seed Oil

Poppy seed oil is edible without refining, as it does not develop rancidity easily. Hot-pressed oil is largely used in soapmaking. It may be rendered edible by refining. The oil yield from black and white seeds is almost similar, but black seeds are more commonly used in expression of the oil because of its easier cultivation. However, white seeds are reported to yield the finest oil. Seeds from the capsules which have not been scarified for opium gave higher yield of oil than those scarified.

Poppy seed is widely used for culinary purposes. It is free from narcotic properties and is used mixed with olive oil, or as a salad oil. It has high digestibility coefficient of about 9% at a daily intake of 50 g. On hydrogenation, it yields a product similar to hydrogenated groundnut oil which may also be useful for industrial purposes.

It is a nonyellowing drying oil forming a hard and lustrous film. The oil is rendered colorless by exposure to sunlight. It dries much faster and more uniformly than linseed oil. Modified oil, with good drying properties for use in paints and varnishes, can be prepared by heating raw oil for 5–12 h at 150–180 degree Celsius in the presence of catalysts. Brown Poppy seed oil on dehydration and treatment with cobalt forms films resistant to boiling water and, in this respect, is equivalent to those formed by dehydrated castor oil.

Table 4.5 Quality analysis of the Poppy seed-cake

Property	Content
Moisture	10.6%
Crude protein	36.6%
N-free extract	20.7%
Ether extract	11.6%
Mineral matter	12.4%
Digestible protein	30.4%
Total digestible nutrient	62.3%
Nutritive ratio	1.0

Poppy seed oil is used in the production of artists' paints; for this purpose, sun-bleached oil from the first cold pressings of the seed is preferred.

The oil is useful in the preparation of linoleic acid, soft soaps, ointments and emulsions, and also applications for skin care. The oil finds use also as an illuminant.

It is used against diarrheas, dysentery and scalds. Lower grades are used for lubrication. Poppy seed oil is used in Europe as a substitute and adulterant to olive oil, with which it resembles very much.

Poppy Seed-Cake

The cake or the material left after extraction of the oil from the Poppy seed is sweet and nutritious and is eaten by poor people. It is readily consumed by cattle and sheep and may be fed alone or preferably mixed with other seeds. However, excessive feeding of the cake (over 1 kg/day) to dairy cattle may decrease in the fat content of the milk and in soft fat of the milk. The cake is not recommended for young or breeding animals.

Analysis of the cake showed the following results tabulated in Table 4.5.

The Poppy seed-cake can also be used as an organic manure.

The Poppy seed-cake may sometimes prove toxic to the cattle, when fed, owing to the presence of alkaloids arising from the contamination of seed with particles of the capsule. This results in marked gastroenteritis, nervous excitement, lack of appetite and colicky pains. Lactation may drastically decrease or totally cease, with the cattle losing weight progressively, leading to fatality, as has been reported in some cases.

Poppy Capsules for Opium Manufacture

Table 4.6 summarizes the opium principle in Poppy capsules.

The morphine variation of 0.18–0.90% refers to analysis of several samples obtained from different countries. The morphine in the capsules decreases rapidly during storage.

Pharmaceutical Uses of Poppy

An infusion of the Poppy capsules is used as a soothing application for bruises and inflammatory swelling and also to cure painful conjunctivitis and inflammation of

Table 4.6 Details of opium principles in Indian Poppy seed

Property	Content
Total alkaloid content	0.4–0.6%
Lanced capsules	0.15–0.22%
Morphine	0.18–0.90%
Nicotine	0.1–0.2%

the ear. A hot decoction of the Poppy capsules is applied as an anodyne. Capsules are also used in the form of a syrup or extract, as a sedative against irritant coughing and sleeplessness. In India, an intoxicating liquor is prepared by heating Poppy capsules with jaggery and water. In the USA, mature Poppy capsules have been processed to yield a liquor which can be used as a substitute for opium in the production of morphine.

Poppy Straw

Dry, uncleaned empty Poppy capsules having a stem of about 7.5 cm have been used in Europe and other places, where the plant is cultivated primarily for its seed and oil as a source of morphine and narcotine. The two alkaloids are obtained in a yield of 0.08 and 0.009, respectively. Poppy straw and heads have been processed to yield concentrated alkaloidal extract, such as "Optopon" (morphine 20–22%; other opium alkaloids 16–18%), which can be used directly as a pharmaceutical product. Poppy straw has also been tried for the manufacture of hand-made boards.

Young Poppy Plant as "Pot Herb" and for Cattle

Young Poppy plant is sometimes eaten like lettuce. It is grown as a pot-herb in Iran and is also fed to cattle. In Europe, instances of poisoning have been reported when cattle have eaten unripe Poppy capsules of the plant.

Leaves and petals have been sued for packing opium. Leaves are also smeared as an anodyne. Leaves contain 0.2–0.3% morphine.

Paper Pulp from Poppy Plant

The Poppy plant has been tried in the production paper pulp. Kraft pulping gave pulps having good strength properties and with usefulness for the manufacture of wrapping papers, bags and other grades requiring an improved formation.

Red Poppy Flowers Used as Pharmaceutical Products and as an Indicator

The Red Poppy flowers are used in medicine for making syrup. The red and lilac flowers contain a coloring matter and are suitable for use as an indicator.

Chapter 5
Star-Anise and Japanese Start-Anise

Abstract *Star-Anise* The chapter would at first discuss the description and distribution of the crop. Crop management would follow. The chapter would lay great emphasis on post-harvest technology of the crop. The manufacturing of essential oil from Star-Anise would be discussed at length. *Japanese Star-Anise* The chapter would discuss the description, distribution crop management, and post-harvest technology of the crop. A discussion on processing technology of the crop also would be incorporated in the chapter.

Keywords for Star-Anise Description · Distribution · Crop management · Star-anise oil · Fatty oil
Keywords for Japanese Star-Anise Description · Distribution · Post-harvest technology processing technology · Uses

Star-Anise

Botanical name: *Illicium verum* Hook.

Family: Magnoliaceae.

Description and Distribution

Star-Anise is the dried, star-shaped fruit derived from *Illicium verum* which is an evergreen tree attaining a height of 8–15 m and a diameter of 25 cm. Leaves are entire, 10–15 cm long, and 2.5–5.0 cm broad, elliptic to oblanceolate; flowers are solitary, white to red colored, fruits are star-shaped, reddish brown in color, consisting of 6–8 capsules (follicles), arranged in a whorl, radiating from the top of a slender short central stem; each follicle is 12–17 mm long, boat-shaped, hard and wrinkled, containing a seed; seeds are brown, compressed-ovoid, smooth, shiny, and brittle. The carpels have a pleasant anise-like odor and an aromatic, refreshing, sweet, anise-like taste; the seeds have the same taste, but to much lesser extent. It is available, whole, and is not usually ground.

The plant is indigenous to tropical and subtropical East Asia. It is extensively cultivated in a limited area with particular ecological factor, in Kwangsi in south-east China, mainland Tonkin in Indochina, and is mostly imported from China and

K. P. Nair, *A Compendium of Unique and Rare Spices*,
https://doi.org/10.1007/978-3-031-20249-0_5

Indochina. No import statistics are, however, available for Star-Anise. It is now produced to some extent in Arunachal Pradesh in north-east India.

Area and Production

Star-Anise is a recent introduction to India, from China, which is the major producer, supplier, and exporter of the crop. No organized commercial-scale cultivation has yet been started in the country, but it is reported to be gradually picking up. According to the latest report by the Deputy Director of The Spices Board, located in Guwahati, in north-east India, about 12.5 tons of Star-Anise is produced annually in Arunachal Pradesh, also in north-east India. No reliable statistics on area are yet available.

Crop Management

A survey of scientific literature on spices did not reveal any first-hand published report from India on Star-Anise, though it is being cultivated in Arunachal Pradesh. Since Star-Anise fruit as well as its essential oil are in considerable demand within and outside the country, the reader is advised to read the report of Guenther (1972).

Soil and Climate

In general, the cultivation of Star-Anise has been limited to a relatively confined area of East Asia. Agricultural Research Stations in other parts of the world have repeatedly attempted to grow the crop, but most of these efforts have failed to yield a commercially worthwhile crop of the little fruit. It would, therefore, appear that ecological factors, such as soil and climate, prevailing in the place like the State of Kwangsi (south-eastern China) and Tonkin (French Indochina) are particularly favorable places for the growth of the crop (Guenther 1972).

Propagation

Propagation of Star-Anise is by seed. Seeds are collected from vigorous old trees, known for their high yield of fruits. Seed is then removed from the fruit with a pen knife; only fully matured seed, which is easily recognized by its brownish color, is retained for propagation. One kilogram of fruit yields about 1000 seeds, suitable for planting. After stratification, seeds are planted, 3–4 cm apart in a well-protected seed bed. Since seed loses its germination power or viability rapidly, it has to be planted within 3 days of harvest. Seed bed preparation is usually started in October–November.

In Tonkin (Indochina), seeds are sown late in the fall season or during the first few months of the year. It would be best to plant seed by June or July, because it has been observed that premature plantings frequently suffer from dryness prevailing in the summer months of April and May.

Raising Nursery and Transplanting

After young plants have developed the fourth leaf, they are transferred to a nursery and the seedlings are planted 25 cm apart. The young trees remain in nursery for 3 years, or until they are sufficiently strong enough, to be transplanted in the main field 5–6 m apart.

Weeding

The trees require no special care, except some weeding around them, in July and August, which is done merely to facilitate gathering of the fruits from the ground level, and to exterminate all straw-like herb plants, liable to burn and spread bush fires.

After-care

1. Thorough plowing and some mulching in the fall provide trees with sufficient moisture during the dry season.
2. Manuring and Fertilization: In a modern Star-Anise plantation, every tree, at the beginning of summer, should receive about 15lbs of stable dung and 150lbs of Ammonium sulfate.

Maturing and Flowering

Star-Anise flowers in an unusual manner. Blossoms appear in the month of March to April end, but, are sterile, and develop no fruits. A second flowering period begins in July or August and lasts about two weeks. The flowers are larger than those of the other periods and develop into fruits prematurely from November to January. The natives call this the "late harvest". The third flowering season starts immediately after the second, in fact, partly dovetailing the second flowering. The flowers at this time are relatively small and develop into fruits very late, usually in August, September, or October of the following year. This, the natives call "the big harvest". The color of the flowers ranges from white to red. The trees flower throughout the year, but, most abundantly, during certain seasons.

Fruit Development

The fruit matures all year round, but, in normal years, the August to October harvest accounts for about 80% of the total harvest.

At the beginning of the fruiting period, the embryonic fruits, developed from the second and third flowering periods of the year, are filiform and do not differ from one another. It is only in October that it becomes possible to differentiate between fruits which will ripen prematurely, from November to January of the same year, and fruit which will ripen from August to October of the following year, when the tree is, again, in full bloom.

Aside from the two harvests described above, there is a less important harvest in March–April. However, because of the strong winds and sudden changes in ambient temperature prevalent in the latter period, most of the fruits on the trees fall off prematurely, after reaching only one-tenth or one-fifth of the normal size. These are especially rich in essential oil (Guenther 1972).

Harvesting of Fruits

Star-Anise fruit contains maximum of essential oil before complete maturity and should be gathered at this stage of development in order to give high yield of oil. Over-ripe naturally developing fruits can be used only for export in the dried state.

In actual practice, the natives, usually the village children, simply climb onto the trees and using long poles with a little hook attached to its end detach the fruit from the branches or shake the branches until the fruits fall to the ground. The branches are sufficiently strong to support harvesters; moreover, the fruits are usually not so far out of reach that the harvester cannot easily detach them.

To prevent breaking the branches and cause damage to the trees, it has been suggested that double ladders be employed or that only the fallen fruits are collected. Neither of these methods, however, is practical. Double ladders cannot be used on trees of such height as that of the Star-Anise crop, whose lowest branches sometimes are 7 feet high above the ground. Nor is it advisable to collect fallen fruits, which, more often than not, could be the damaged ones.

Yield of Fruit

A Star-Anise planting comes into production when about 10 years old. The first harvest is small, amounting to only 0.5–1.0 kg of fruits per tree. In the 15th year since planting, a tree yields, approximately, 20 kg and the planting turns profitable. Full production sets in after the 20th year since planting, with a yield of about 30 kg of fruits per tree. The average lifespan of a tree is 80–100 years, or, even more. The productivity of a Star-Anise tree like that of most nontrimmed fruit trees varies considerably from year to year. A tree yielding 30 kg of fruits in one year may yield only 1 kg in the following year.

According to one report, the productivity of a Star-Anise tree may be estimated as follows: A tree 13–25 years old may yield from 0 to 5 kg of fresh fruits per tree in poor years and from 5 to 10 kg in normal years. A tree older than 20 years may yield from 0 to 10 kg fruits in poor years and from 10 to 20 kg fruits per tree in normal years and from 20 to 40 kg per tree in very good years. All these figures apply to fresh fruits (Guenther 1972).

According to yet another report, a worthwhile harvest cannot be expected before the 10th year since planting, and, from the 10th to the 20th year, a tree yields 30–35 kg of fruits (which yields about 600 g of oil). From the 20th year onwards, a tree produces, on average, 40–50 kg of fruits (Guenther 1972).

Post-harvest Technology

Drying of Fruit

On drying in flat baskets, by exposure to the sun for about 10 days, 100 kg of fresh fruits (green) will yield 25–30 kg dried fruits (The ratio is approximately 3:1—one third). To preserve them better, the Chinese frequently boil the fruits for a few minutes before drying; however, they lose part of the essential oil. The dried fruit is exported as such and represents the well-known brown Star-Anise of the drug market.

Processing Technology and End Products

Manufacture of Essential Oil

The fruits are mostly distilled fresh (green). If too many fruits have accumulated at a distillation post, they may be kept for about 10 days, or even longer, provided they are spread out in a thin layer and frequently turned over, to prevent fermentation. Occasionally, the demand for dried fruits for export slackens and prices then fall to a very low level. Whenever this happens, the natives use their accumulated stocks of dried fruits for distillation, hoping to obtain a better return from the essential oil obtained after distillation.

Production of the essential oil is entirely in native hands. The country stills used for the purpose resemble those employed for distillation of Cassia oil in China. They are primitive, although with quite clear construction. They hold up to 300 kg of fresh fruits per charge. In some regions, the natives break fruits by hand, prior to distillation. The fruits (in most stills about 180 kg) are placed in the retort, together with sufficient water to cover the material. The heating by direct fire beneath the retort vats is to be done slowly to prevent boiling over of the charge and escape of vapors through loose-fitting joints of the system. Steam and oil vapors pass from the retort through three small holes at the bottom of the vase-shaped still-head, which also serves as a sort of condenser. The still-head is covered with a flat bowl, through which cold water flows. Steam and oil vapors are condensed on the bottom of the flat bowl, the condensate drips into the lower part of the still-head, and from there, through a pipe, into the oil separator. Here, the oil and water separate, and the distillation water flows back into the retort and is re-distilled (cohobated). Because of the small condenser surface, distillation in these stills must be carried out slowly. A charge of fresh fruits requires 48 h, and a charge of dried fruits requires 60 h.

Native Versus Modern Distillation of Oil

Though it is primitive, production of Star-Anise oil in small native distillation posts offers, nevertheless, one great advantage and that is the cost factor. It is low cost. The trees planted many years ago grow semi-wild near the villages and require no care. Harvesting carried out by village children costs next to nothing. Any investment in simple stills was amortized long ago. Distillation is a mere side operation of villagers; it goes along, at a leisurely pace, with their principal task, which is growing rice. The male members of the family tend to their rice plots, casting an occasional glance at the flow of the distillate in their stills, and separating the oil; the women cook food near the still and kindle fire beneath. In other words, the oil is produced at practically no cost, and this simply means additional income for the family.

Modern Distillery for Oil Merits and Demerits

The erection of a modern distillery would be, relatively, a costly enterprise. Building and stills would have to be amortized, foreman and laborers to be hired, at regular wages, fuel purchased, and local taxes would have to be paid. True, in modern steam stills, the time of distillation can be shortened, substantially, but the savings in the cost of fuel would not compensate for other expenses involved. The only advantage would be in the production of a very high grade oil, free of all adulterants and other unwanted and extraneous materials. Such oil would consist almost entirely of Anethole and possess an exceptionally fine odor and flavor. The cheapest way to produce an oil of very high quality is simply to submit native-distilled oils for rectification, a practice,

Table 5.1 The chemical characteristics of Star-Anise oil

Property	Content
Specific gravity at 25 °C	0. 978–0.0.987
Refractive index at 20 °C	1.5530–1.5582
Optical rotation	1.77–0.57°
Congealing point	15.0–18.4°
Sol. in	1.0–2.5 vol. of 90% alcohol

Anethole is the chief constituent (85–90%)

which was followed by a large European dealer of Star-Anise oil in Langson, Tonkin, some years ago.

Yield of Oil

It varies from 3 to 3.5% of fresh fruits (Pruthi 1998). It is difficult to state exactly yield figures, because this varies according to the maturity of the fruit, location where the tree is grown, and condition of the tree, region where it is grown and climatic conditions. According to one report, yields ranged from 3.1 to 3.5% of fresh fruits from Langson and Diem-Her and 2.2–2.5% of the material from That Khe and Loc Binh.

Elberhardt (1927) reported that 180 kg of broken fruits, on distillation for 45 h, yield approximately 5.5 kg of oil (about 3%).

In general, 100 kg of fresh fruits, distilled for 48 h, yield 2.5–3.0 kg of oil; 100 kg of dried fruits, distilled for 60 h, gave 8–9 kg of oil.

With codistillation, dried Star-Anise with glycol, *in vacuum*, Sabetay (1940) obtained 8.25% of volatile oil.

Quality of Star-Anise Oil

It is colorless or pale yellow with characteristic odor and aromatic taste of true Anise oil (from *Pimpinella anisum*); the odor and taste of the latter are rather more delicate. The oil has the following characteristics, as tabulated in Table 5.1.

Star-Anise oil is official in the pharmacopoeia of many countries. It constitutes the bulk of the oil of Anise of commerce (Pruthi 1998).

Oil of acceptable quality should have a congealing point above 15°, a specific gravity above 0.978 at 25 °C and a refractive index above 1.5330 (Guenther 1972).

The quality of Star-Anise oil, like that of Aniseed oil, can be evaluated by its congealing point. In fact, it is possible even to estimate Anethole content, the major constituent of the oil (85–90%) directly from its congealing point.

Quality Grading of Star-Anise Oil

Table 5.2 summarizes the oils graded according to their congealing point.

Oil having a congealing point below +15 °C is not acceptable in international trade. The congealing point of Star-Anise oils and, hence, their Anethole content depend chiefly on two factors, namely season in which the crop is grown (In cold season, it is +16 °C or above, but in summer season it is around +15 °C) and

Table 5.2 Oil classification according to congealing point

Quality grade	Congeaing point (+ °C)
Best	18
Very good	17
Good	16
Fair (lowest limit)	15

composition of distillation material; addition of leaves and branches definitely lowers Anethole content and hence the congealing point of commercial quality, since the congealing point of leaves averages above 13°. This constitutes an adulteration of Star-Anise oil.

Leaf Oil

The leaves of Star-Anise tree yield about 0.5% essential oil on steam distillation, which is inferior to that of the fruit, both with respect to low congealing pint (average 13%) and aroma.

Adulteration

Star-Anise fruit is often adulterated with fruit of *I. anisatum* Linn.syn. *I. religiosum* Sieb. & Zucc., grown in Japan, and known as "Japanese Sacred Anise Tree", "Poison Bay", "Shikimi", or "Shikiminoki", "False Aniseed", and "Bastard Star-Anise". The fruit of *I. anisatum* is believed to be poisonous in Japan.

The Chinese Star-Anise oil is adulterated with its branch and leaf oils.

Fatty oil

Decorticated Star-Anise seeds contain a fatty oil (55%) having the following constants, as shown in Table 5.3.

Uses

Star-Anise as A Flavorant

Star-Anise fruit has an agreeable, aromatic, sweet taste, and a pleasant odor, resembling Anise. It is used as a condiment to flavor curries, confectioneries, and spirits, and for pickling. It is also used in perfumery. The fruit is chewed to sweeten breath and to help digestion.

As a Pharmaceutical Product

It is stomachic and carminative and considered useful in controlling flatulence, spasmodic affection of the intestinal canal and dysentery. It is used as an adjunct to cough mixtures and as a corrective of taste. It is also used in production of Absinthe (spirit) (Pruthi 1998).

Table 5.3 The chemical constants of decorticated fatty oil of Star-Anise

Property	Constant
Specific gravity at 25 °C	0.9128
Refractive Index at 25 °C	1.4677
Saponification value	194.5
Iodine value	88.36
Acid value	11.65
Acetyl value	8.37
R.M. value	0.75
Polenske value	0.29
Unsaponification matter	0.59
Components of the fatty oil are (%)	
Myristic acid	4.43
Stearic acid	7.93
Oleic acid	63.2
Linoleic acid	24.4

Star-Anise Oil

Star-Anise oil is used to flavor confectionery, candy, chewing gum, tobacco, animal feeds, and liquors and in other pharmaceutical preparations. It is also used in the manufacture of perfumes and soaps.

The oil is a stimulant, stomachic, carminative, mildly expectorant, and diuretic. It relieves colic and is an ingredient of cough lozenges. The oil is employed as an application in rheumatism and otalgia and as an antiseptic. It is considered useful against body lice and bed bugs and forms an ingredient of cattle sprays. It is used in favus (honeycomb ringworm) and scabies. The oil enters into the composition of a number of official preparations.

Star-Anise paste is used to flavor foods and confectionery and is considered useful for colic, against constipation and insomnia. The Chinese are reported to prepare a medicinal tea from leaves of the Star-Anise tree, and they claim that one or two carpels of Star-Anise when added to chicken, which is to be roasted, improves its flavor remarkably (Pruthi 1998).

Japanese Star-Anise

Botanical name: *Illicium religiosum* Sieb et Zucc.

Family: Magnoliaceae.

Description and Distribution

The Japanese Star-Anise *Illicium religiosum* Sieb et Zucc (family Magnoliaceae), known as "*Shikinami*" in Japan, is an evergreen tree with a trunk of about 3 m high, with pale, yellowish, white blossoms, which bloom from spring to summer. The tree grows wild in warm localities of southern and central Japan, on the Loochoo islands and in Formosa. It is grown extensively in the Prefecture of Nagasaki (chiefly on Goto Island) and, to a smaller extent, in the Prefectures of Kochi and Tokushima on the island of Shikoku.

For a long time, the Japanese have been planting Star-Anise tree in temple compounds and in cemeteries in order to protect them from desecration by wild animals. This habit appears to have developed from the fact that the fruit of the tree is poisonous and that leaves emanate a peculiar odor which is supposed to keep the animals at bay. The custom has developed to such an extent that even today altars during funeral services are decorated with leaves of *Illicium religiosum*.

Area, Production, and Exports

No recent reliable statistics are, yet, available on area and production. As per one report, the total harvest of Star-Anise in Japan is 200,000–500,000 kin per year (1 kin = 0.6 kg or totally 120–300 tons). And, only about 10% of this is used domestically, 90% being exported. Prior to World War II, about 60% of the exports were to India for use as incense in religious ceremonies, and 40% was shipped to China for admixture with Chinese products and consumption as a spice in cooking. From the end of the hostilities to the end of 1949, about 60,000 kg of Japanese Star-Anise fruits were imported into the United States of America for unknown purposes.

Crop Management

No precise information is available on its cultural aspects. However, it is reported that they are practically the same as those for the Chinese Star-Anise.

Post-harvest Technology

The fruits of the Japanese Star-Anise tree bear a striking resemblance to that of the Chinese tree (*Illicium verum* Hook. f.) excepting that the former is much smaller. They are harvested and sun-dried like the Chinese Star-Anise.

Processing Technology

The dried fruits of *Illicium religiosum* contain about 1% of volatile oil with an unpleasant odor, quite different from that of the Chinese product (*Illicium verum* Hook f.). The oil is not produced commercially, yet. It is reported to be toxic. Its toxicity has been found to be due to the occurrence of a poisonous principle or a compound identified as Henanomin. Its chemical composition is quite different from that of the Chinese Star-Anise and has been reviewed by Guenther (1972).

Use

In Japan, the fruit is gathered for use in the manufacture of incense sticks and for consumption as a spice.

It has stomachic and tonic properties, but, owing to its toxicity, it can be consumed only in very small quantities. Some years ago, a shipment of the Chinese Star-Anise, which contained admixed Japanese Star-Anise, was exported to Germany, and many Germans suffered from poisoning in Hamburg (Guenther 1972).

References

Elberhardt (1927) Bull.agence gen.colontes parts No. 447
Guenther E (1972) The essential oils, 5th print. Van Nostrand Reinhold Co., New York
Misaki (1932) Aichi Igakuka 39:1123
Pruthi JS (1998) Spices and condiments, 5th edn. National Book Trust, New Delhi, India

Chapter 6
Sage

Abstract The chapter would at first discuss the description and distribution of sage. Details on area, production, and trade would follow. The chapter would also include a discussion on the major types of sage in world trade. Further discussion would be on crop management and post-harvest technology.

Keywords Description · Distribution · Processed products · Volatile oil

Botanical name: *Salvia officinalis* Linn.

Family: Labiatae

Description and Distribution

Sage is the dried leaf of the plant *Salvia officinalis* Linn. Which is a member of the mint family. It grows and is also cultivated in Yugoslavia, Italy, Albania, Turkey, Portugal, Spain, Cyprus, England, Canada, and the USA. It is also grown in several other European countries. It is native of the Mediterranean region.

It is a hardy and a variable subshrub and is often cultivated as a spice and for medicinal purposes. Stems are shrubby, white and woolly, 30-cm tall; the leaves are grayish green, aromatic, entire, petiolate, oblong, 7–8 cm long, base narrowed or spear shaped or round. On drying, the leaves turn silvery gray, and they have soft velvety texture. The flowers are blue, purple, or white in simple racemes.

The cultivation of Sage or "Garden Sage" has been tried for its essential oil in Jammu (Jammu and Kashmir State of India), which is currently being imported for use in perfumery and soap industries. The plants are reported to have done quite well. It should be propagated in India to save foreign exchange.

The cultivation of Sage in the Dalmatian region of Yugoslavia is considered by the trade to be the best in quality, and the product derived from the cultivated plants in this region is rated the next in preference. The odor is strong, fragrant, and aromatic, and the taste is aromatic, warm, somewhat astringent, and a little bitter. Other product such as Spanish Sage and Greek sage derived from the other species of *Salvia* are used either as a substitute or adulterant of the genuine product.

K. P. Nair, *A Compendium of Unique and Rare Spices*,
https://doi.org/10.1007/978-3-031-20249-0_6

Area, Production and Trade

No reliable published information is available on both area and production of Sage in India or in the world, at large. However, the available information on the international trade of Sage is summarized as follows

Albania exported 867 tons followed by Yugoslavia at 443 tons, followed by Turkey at 28 tons and Italy at 23 tons, and lastly USA, the exact trade figure being unknown. West Germany and Mexico also exported Sage to the USA. During 1977, the USA imported 1393.4 tons of Sage valued at US $1.620 million. Hence, there is very good scope for its cultivation in India, especially for export.

Major Types of Sage in World Trade

Dalmatian Sage

This is the term for the product from Yugoslavia, derived from the Dalmatian Coast. This has been traditionally the prime source of Sage, and its product is the type which commands the highest price. It is highly aromatic, with a pleasing, in particular, mellow flavor—the mellowness being the hallmark among the different varieties of Sage. The Dalmatian sage is shipped in tightly compressed bales approximately 110 pounds each.

Albanian Sage

Albania is the other State for the Sage. This is also recognized as an excellent Sage, but does not typically draw the same demand as the Dalmatian Sage in the USA. While the two are comparable in the content of the essential oil, there are certain differences in the flavor characteristics, due to the variance in the components of the oil. These components gave the Dalmatian Sage its smoother, more fully rounded flavor notes. Albanian sage is shipped in loose semi-pressed bales.

Other Types

In any given year, there are several other types of Sage, known by their national origins, which come to the US market. Turkish, Italian, Spanish and Cretan are typical. But, there are other sources, and they vary in trade volume from year to year. In total, the poundage of all these other types is relatively small, as compared to the shipments of Dalmatian and Albanian Sage. The other varieties contain flavor notes which are harsher than those of *Salvia officinalis Linn.* Sage.

Different Forms of Sage

For the industrial user, Sage leaf is available in the market in a variety of forms, namely (1) Whole, (2) Cut, (3) Rubbed and (4) Further Ground to Specified Granulations.

Whole Leaf Sage

This is the term for the product which has been cleaned and prepared for sale, but, not cut, or otherwise processed.

Cut sage

This is known variously as (1) Cut, (2) Cracked, (3) Sliced, (4) Dropped, (5) Butcher's Chop, etc. Whatever the term, it refers to leaves which have been cut (rather than ground) into smaller pieces—usually to the buyer's specification. Particle sizes of 0.3–0.6 cm are common, but size is largely up to the buyer. Cut Sage is the choice where the manufacturer wants the sage to be apparent in the end product.

Rubbed Sage

This is the traditional term for Sage which has been put through only a minimum grinding and a coarse sieve. The result, due to the soft, velvety texture of the Sage leaves, is a fluffy, almost cotton-like product, unique among ground herbs. Many sausage makers prefer this minimal processing for Sage in the opinion that it preserves flavor longer and at the same time makes it suitable for easy blending into a product. More Sage is sold in the rubbed form than in any other. However, the buyer should be aware that "Rubbed" is the traditional term, not a standardized one. The consistency can vary according to the processing procedure used by the supplier.

Ground Sage

This is often called "Finely Ground" to differentiate it from the "Rubbed", which also goes through a grinding process. In general, this simply means that the Sage has been further pulverized and sifted through finer sieves. The final granulation (particle—size) is up to the buyer and this, in turn, to the end-user. Basically, the finer the powder, the more completely the Sage will blend into the food, being made, and lesser are its chances to be seen in the finished product. This can be important in some products, especially light-colored ones and those in which Sage is not meant to be for noticeable characteristics.

Crop Management

The cultivation of garden Sage has been tried in Jammu, India, for its essential oil, which at present is being imported into India or the manufacture of soap and perfumery industries. The plants are reported to have performed well; the quality of the oil is comparable with oils of overseas origin.

Climate and Soil Requirements

A hot and dry climate is not suitable for the cultivation of Sage, since it produces an inferior crop. Sage is said to thrive best on any soil that grows cotton, but a product of better quality can be obtained on rich clayey and loamy soils. The crop can be grown as an annual or perennial.

Propagation

Sage plantations are preferably raised from rooted stem-cuttings, as the crop raised from seeds is likely to be a mixture of different types. The cuttings are obtained preferably from broad-leaved plants which have the least tendency to flower.

Planting

The rooted cuttings are planted 30–45 cm apart in rows 0.9 m apart. One kg of seeds provides about 20,000–30,000 seedlings, enough to plant one hectare. Seeds are drilled directly in the field, or seedlings are raised in the nursery for transplantation subsequently.

Harvesting

Young plants which have not reached flowering or seedling stage possess finest aroma and contain highest amount of oil. At this sage, harvesting is done by hand or by cutting tops with a mower. Sometimes, the whole plants are cut by small sickles and dried in the shade outdoors or inside the warehouses where warm air is circulated.

Yield

The first-year crop is usually light, but, after the plants are established, 2 or more cuttings in one year may be possible. In the second and subsequent years, a crop of 1750–2250 kg of herbage per hectare may be obtained.

Sage plantations should be renewed every 4–5 years, since by this time plants are likely to become leggy and show signs of die-back, and plantations may tend to become weedy.

Post-harvest Technology

Drying

Sage is dried in the shade in order to retain as much of the natural color and flavor as possible. The leaves and small tops are tied into small bundles or spread on screens and dried in a well-ventilated warm room, away from direct sunlight. If the leaves are dusty or gritty, they are washed in cold water before drying.

The dried bundles can be sold without further treatment or the leaves may be cut, rubbed or pulverized for the packeted herb trade.

Table 6.1 gives the physico-chemical composition of Sage

Adulteration

Other products, such as Spanish Sage and Greek Sage derived from other species of *Salvia officinalis* Linn., are used either as substitutes or adulterants of the genuine product. In the USA, impurities in the manufactured products in trade have been limited to 10% of the stems of the plant and not more than 2% of the other organic matter.

Marketing/Trading

In general, the buyer should look for fresh and rich aroma in Sage, and flavor quality consistent with type (as discussed). Color is silvery gray, but it should be obviously fresh in appearance, not at all faded.

As with all spices, the best guarantee of "choice product" and "consistent quality" in Sage is the use of a reputable, well-established spice-grading company as the source of supply. These firms have the equipment and quality control program to clean and

Table 6.1 Physico-chemical characteristics of Sage leaf

Property	Content
Moisture	5.7%
Protein	10.2%
Fat	14.1%
Crude fiber	16.0%
Carbohydrates	46.3%
Total Ash	7.7%
Calcium	1.8%
Phosphorus	0.09%
Iron	0.03%
Sodium	0.01%
Potassium	1.0%
Vitamin A	23,951 (IU—International Units)/100 g
Vitamin B_1	0.75 mg/100 g
Vitamin B_2	0.34.8 mg/100 g
Niacin	5.7 mg/100 g
Calorific value	415 calories/100 g

process spices to buyers' specifications and to deliver the same quality on each order. If desired, such companies can help buyer to develop realistic specifications. By basing these specifications on the standardized procedures, the buyer and the seller gain a common language, making specifications clear and more workable.

How to Handle and Store Sage

Şage should be stored under cool, dry conditions. Excessive heat will rob it of the flavor and dampness will tend to cake the powdered product, in particular. As with any highly aromatic herb, it should be further protected against flavor loss by making sure that the container of the produce is always tightly closed after every use. Under good storage practices, it will retain its flavor qualities and strength long enough to meet any normal food-manufacturing requirements. Normally, its storage life can be from 9 to 12 months at room temperature.

Processed Products

Volatile Oil

On steam distillation, dry leaves yield an essential oil, 1.3–2.6% by weight of the dry leaves distilled. The oil is produced commercially in Yugoslavia, Spain, Corfu, Syria and the former USSR, and the samples from different sources show considerable variation in properties and constituents. Table 6.2 summarizes these data.

Table 6.2 The chemical properties of Sage essential oil

Property	Content
Specific gravity	0.9268
Refractive index	1.4633
Optical rotation	0.2°
Acid value	1.1
Ester value after acetylation	30.0

The constituents reported to be present are as follows: α, pinene (1.8%), cineole, linalyl acetate (10.1%), thujone (44.45%), borneol, bornyl acetate, farnesol and camphor

Note The above data refer to an analysis of an oil sample obtained from Kashmir by hydro-distillation of the dried herb (yield 1.1%), and linalyl acetate content of the oil is too low for commercial exploitation of the oil for its ester content. The quality of Sage oil is determined on the basis of its thujone content. The higher the thujone content, the better the quality of the oil

In addition to the essential oil, Sage leaves also contain tannin (3%), fumaric, malic and ursolic acids, a bitter principle- picrosalvin, saponin, pentoses, a wax and potassium nitrate.

Uses

As A Food Flavorant

Sage is the most popular herb for use in culinary preparations in the Western hemisphere. It is used in preparing pork sausage and baked loaf. The color of the dried herb is gray, tinged with green; the aroma is strong, fragrant and spicy. The taste is fragrant, spicy, warm, astringent and a little bitter. Sage is the most important herb in every kitchen for flavoring meat and fish dishes and in the preparation of poultry stuffings. It is used largely in poultry dressing, sausage, liver sausage and hamburger seasoning in the western countries.

Sage has been extensively employed in the food industry and as a standard spice in making stuffing for fowl, meats and sausages. It is one of the most important culinary herbs. Dried and powdered leaves are mixed with cooked vegetables and sprinkled on cheese dishes, cooked meats and other similar preparations. Fresh Sage leaves are used in salads and sandwiches. The young leaves are pickled and used for making tea. The apples or tumors on the Sage plant caused as a result of puncturing by a wasp species of *Cynips-Cynips quercusfolii*, a gall wasp species in the genus *Cynips*, are made into a conserve with honey.

In Pharmaceutical Products

Sage is used as a mild tonic, astringent and carminative. An infusion of its leaves is used as a gargle in the treatment of sore throat; hot infusion is said to be diaphoretic. Extracts of Sage leaves are also reported to be antipyretic. A strong infusion of the herb is used to dry up the breast milk for weaning children. Sage has been prescribed to cure female disorders since ancient times, and estrogenic substances are said to have been extracted from the dried leafy tops. Dried leaves are used in tooth and mouth washes, gargles, poultices, tooth powders, hair tonics and hair dressings.

Use in Perfumery

Sage oil finds use in the manufacture of perfumes as a deodorant, in insecticidal preparations, for the treatment of thrush and gingivitis and as a carminative. The oil is used as a convulsant, and it resembles worm wood oil in action, but is less active.

Sage oil is used to adulterate Rosemary and Lavender oils. Sage oil itself is adulterated with American cedar leaf oil, which also contains thujone.

As An Antioxidant

Sage and Sage oil exhibit antioxidant properties. Five antioxidant fractions with antioxidant indices between 8.8 and 10.0 have been isolated from the leaves of *Salvia officinalis*, one of which appears to be poly-hydric phenol.

Use of Sage Residue

After steam distillation, the residual plant material still contains constituents of considerable flavor value. Hot-water extracts of the material on concentration yield an oleoresin, which is used in conjunction with the oil to flavor foods.

Use of Sage Seeds

Sage seed contains 18% protein and yield a drying oil, which is used as a bonding agent in the production of oil plants.

Clary Sage

Introduction

Clary Sage (*Salvia sclarea* Linn.) is an exotic plant and was introduced into India long ago as an ornamental plant in the gardens. This is commonly known as "Garden Sage" or Clary Sage. *Salvia sclarea* Linn was introduced at the All India Coordinated Research Project (M&AP) at the Solan Center, in India, during 1976 from Bulgaria through NBPGR (National Bureau of Plant Genetic Resources), New Delhi, India, which has established well in the country, especially under mid-hills and low valley of the State of Himachal Pradesh.

It is an annual plant with large, broad, hairy leaves. It bears, a long, flowering shoot with light purple flowers. The flowering shoots on steam distillation yield a commercially important essential oil with typical fragrance, which blends well with Lavender, Bergamot and Jasmine oils. It has also found a place in cosmetic industry, flavoring alcoholic and nonalcoholic beverages (Guenther 1974; Hussain et al. 1998).Therefore, keeping in view its multifarious uses and the market demand, there is an urgent need to develop its cultivation practices so that its oil production may be increased and imports are minimized.

Clary Sage, being a herbaceous crop, responds favorably well to higher application of nitrogenous fertilizers. Another fact to be considered in its cultivation is the plant density per hectare. The plant density influences both the growth of the plant as well as its final yield. Thus, the nature of interaction of these two factors needs thorough investigation. Since the information is not available for mid-hills and valley areas of Himachal Pradesh, Rastogi et al. (1997) have conducted systematic investigations on the response of Clary Sage to both Nitrogen fertilizer application and Plant density. The results are briefly summarized below:

For Clary Sage, an application rate of 90 kg N/ha gave maximum flowering shoot yield (60.52 quintal/ha, 1 quintal = 100 kg). The essential oil yield was 38.7 l/ha. This rate of N fertilizer application showed statistically significant higher yield rates (flowering shoot and oil yield) as compared to all other rates of N fertilizer application tried in this investigation. A spacing of 30 cm × 45 cm (plant to plant and row to row, respectively) gave maximum flowering shoot and oil yields. Wider spacing resulted in lowering of yields of both attributes. Yield attributes followed the trend similar to that of the maximum flowering yield, as well.

References

Guenther E (1974) The essential oils. Robert E. Krieger Pb. Co. New York, p 719

Hussain A et al (1998) Clary sage oil. In: Major essential; oil beating plants of India. CIMAP, Lucknow, India, pp 58–61

Rastogi JMS et al (1997) Response of clary sage (*Salvia sclarea* Linn.) to nitrogen levels and row spacing. Indian Perfumer 41(3):113–116

Chapter 7
Savory and Tarragon

Abstract *Savory* The chapter would, at first, discuss the description and distribution of the crop, which would be followed by a discussion on crop management, post-harvest technology and processing technology; the latter would include a discussion on essential oil of savory. *Tarragon* At first, the chapter would discuss the description and distribution of the crop, followed by a discussion on crop management, post-harvest technology, processing technology and essential oil of tarragon.

Keywords for Savory Description · Distribution · Crop management · Post-harvest technology · Processing technology · Essential oil
Keywords for Tarragon Description · Distribution · Crop management · Post-harvest technology · Processing technology · Essential oil

Savory

Botanical name: *Satureja hortensis* Linn.

Family: Labiatae

Description and Distribution

There are two types of Savory as follows:

1. *Satureja hortensis* Linn.
2. *Satureja montana* Linn.

Both possess a pronounced thyme-like odor and flavor. And, both are employed in culinary preparations. The oil of the two plants is closely related in chemical composition. The oil can be obtained by steam distillation of the herb.

The dried leaves and flowering tops derived from *S.hortensis* constitute winter Savory of commerce. The best quality of savory comprises of dried leaves; only. *S.hortensis* constitutes summer Savory. Savory is an erect pubescent annual herb, 20–30 cm in height with pinkish branches, fund in Kashmir State of India. Leaves are oblong—linear or lanceolate with deep-pitted glands on both sides. Flowers are small axillary cymes. It grows in Southern France, Germany, Spain and other parts of Europe, also in England, Canada and the USA. The dried leaves are brownish green in color, up to 10 mm in length. The odor is strong, warm and highly aromatic;

K. P. Nair, *A Compendium of Unique and Rare Spices*,
https://doi.org/10.1007/978-3-031-20249-0_7

the taste is somewhat sharp and camphoraceous. It is marketed both in whole and ground form.

Satureja hortensis Linn., the summer Savory or Garden type, is largely cultivated, although it also grows wild, to some extent. An annual, it is less developed than *Satureja montana*. Its stalks are herbaceous, covered with soft and linear leaves. The branches frequently are reddish in color. Flowers of the Garden Sage are whitish, dotted with red, growing in a bouquet from the summit of the branches.

Satureja montana Linn., the winter Savory, as the French call it "Poivre Diane" grows wild on sunny and calcareous and hill side of Southern France, especially in the Alpes-Maritimes. It is a hardy, shrub-like perennial, attaining a height of 8–14 in. with small, yellowish or slightly purple leaves, growing alone on top of the branches. The leaves are tough and linear, the stalks are woody at the base, and the seeds are brown and finely shagreened. The plant seeds itself.

Area, Production and Exports

In India, it grows naturally, and no commercial-scale cultivation has, yet, been undertaken. Hence, no statistics are available on these aspects. *Satureja montana* is rarely cultivated overseas.

Crop Management

In the absence of any Indian published reports on the cultivation of Savory or the cultural aspects, the available information from overseas, on these aspects, is summarized below for the benefit of the entrepreneurs/farmers in the cultivation and tapping the economic potential of Savory

Soil and Climate

Savory needs light-textured soil

Propagation

Propagation is normally by seed of *Satureja montana*. It is by seed or root divisions, vegetatively.

Seed Rate, Method of Sowing and Weeding

The seed is sown manually in light-textured soil, at the end of April in Northern France and at the end of February-March in Southern France. It begins to sprout after about a week or slightly longer.

The seed should be sown at a rate of 3–4 kg/acre in rows about 12 in. apart. The seed sprouts very slowly and requires intense culture and two hand-weedings.

Harvest

Savory is ready for harvest when flowers begin to bloom.

Yield

About 6 US tons per acre of clover-dried herb gives 26 US tons of oil (8.2%).

Table 7.1 The physico-chemical properties of the dried Savory herb

Property	Content
Moisture	9.1%
Protein	7.1%
Fat	5.2%
Crude fiber	15.3%
Carbohydrates	54.6%
Total ash	8.7%
Calcium	2.2%
Phosphorus	0.14%
Iron	0.04%
Sodium	0.02%
Potassium	1.1%
Vitamin A	865 IU (IU = International Units mg/100 g)
Vitamin B_1	0.37 mg/100 g
Vitamin B_2	0.04 mg/100 g
Vitamin C (ascorbic acid)	12.0
Niacin	4.1 mg/100 g
Calorific value	355 cal/100 g

Post-harvest Technology

The flowering tops and leaves are gently dried in shade for later sale to herbalist's trade. It is marketed as both "whole" or "ground" forms.

Processing Technology

Essential oil of Savory

In both summer and winter Savory crops, fresh flowering tops and leaves are used. The yield of volatile oil varies from 0.1 to 0.2% from fresh leaves and flowering tops.

The plant material is charged into a still. Steam generated in a separate steam generator is injected into the plant material, and the oil is separated in a Florentine flask.

Table 7.1 summarizes the physico-chemical composition of the dried Savory herb.

Composition of Fresh Herb

Table 7.2 summarizes the composition of fresh green herb.

The dry matter of fresh green herb contains 11.95% pentosans.

Volatile Oil

Table 7.3 summarizes the physico-chemical properties of the volatile oil

The recovery of essential oil from the herb is higher during full flowering.

Table 7.2 Composition of fresh green herb

Property	Content (%)
Moisture	71.88
Protein	4.15
Fat	1.65
Sugar	2.45
Crude fiber	8.6
Total ash	2.11

Table 7.3 Physico-chemical properties of the volatile oil

Property	Content
Specific gravity at 15 °C	0.896–0.960
Optical rotation	−0 0.56° +0 0 4°
Phenols	30–42% and soluble in 2 volumes of 80% alcohol

Adulteration of Essential Oil

Adulteration of essential oil of Savory is frequent, with much cheaper priced oil of Origanum, especially in view of the fact that it can hardly be detected by chemical analysis. The chemical constants of Savory and Origanum oils are similar. However, an expert familiar with the delicate and characteristic odor and flavor of genuine Savory oil will only have little difficulty recognizing sophistication with Origanum oil which possesses a coarse, phenol-like character. This is also true if the Savory oil has been adulterated with synthetic carvacrol.

Occasionally, Thyme oil (*Thymus vulgaris* Linn.) is used for adulteration of Savory oil, both oils possessing similar properties. However, Thyme oil contains thymol, a crystallizable phenol as the main constituent which can be separated by congealing, whereas Savory oil contains carvacrol, as the main chemical constituent, a noncrystallizable phenol.

Uses

Savory oil is used in flavoring soups, sauces, eggs, salad dishes, canned meats and poultry dressing. It is used in pork sausage, and vegetables take on a rare new flavor when tiny leaves of summer Savory are used instead of Parsley or Chives. Savory oil also finds use in most of the food items listed above.

The herb is reported to be carminative and stimulating. The antioxidant property of the herb reported, in part, is due to the labiatic acid contained in Savory leaves.

Tarragon

Botanical name: *Artemisia dracunculus*. Linn.

Family: Asteraceae

Description and Distribution

The dried leaves and flowering tops of *Artemisia dracunculus* Linn. constitute what is known as "Tarragon", or, "Estragon", so well known to connoisseurs, the world over, for its extraordinary flavor. Sometimes, it is also called French Tarragon. Its aroma is warm, aromatic and reminiscent of Anise. It is a perennial herb found in Western Tibet (4265–4880 m above mean sea level) and in Lahaul. It is distributed in Afghanistan, Western Asia and South and Middle Russia. It is cultivated in Southern Europe (France, Spain), temperate zones in the USA, as well as throughout the colder New England area. It grows to a height of about 2 ft (Pruthi 1997).

Area, Production, Exports

No information is currently available.

Crop Management

Soil and Climate

Deel and Deel (1929) found that optimum yield of the herb material and essential oil are obtained in soil with a pH of 6.2 (in such a soil, the oil may contain as much as 66.3% of methyl chavicol, its chief chemical constituent).

Propagation

Through root subdivision/segments

Planting Season

For planting, a cluster of Estragon roots is divided into several segments, and these segments are planted in the field in the months of February/March.

Harvest

The first harvest takes place in July. The second is in September. The plants are cut in a day, flowers and leaves together, and only very short stumps are left above ground. These stumps grow sufficiently to form second (September) harvest. One planting lasts about three years; after which, it must be renewed. An important precaution in harvesting to be taken is that the plant should be cut in the beginning of the flowering stage; otherwise, it loses some of its aroma and the essential oil content.

Post-harvest Technology

Flower tops and leaves are dried in shadow to "clover dry" condition. Excessive drying lowers the yield.

Composition

Table 7.4 summarizes the physico-chemical composition of dried leaves of Tarragon.

Table 7.4 The physico-chemical composition of Tarragon dried leaves

Property	Content
Moisture	4.7%
Protein	24.2%
Fat	7.3%
Crude fiber	6.8%
Carbohydrates	4.7%
Mineral matter	12.3%
Calcium	1.3%
Phosphorus	0.31%
Iron	0.034%
Sodium	0.07%
Potassium	3.2%
Vitamin A	175 IU (international units)/100 g
Vitamin B_1	0.2 mg/100 g
Vitamin B_2	1.34 mg/100 g
Niacin	8.9
Vitamin C (ascorbic acid)	12.0 mg/100 g
Calorific value	365 cal/100 g

Processing Technology

Essential Oil

Distillation of herb is carried out in direct steam—stills. 1–1.5 h are required to complete a charge. Care must be exercised during distillation not to burn the plant material in distillation stills since this adversely affects odor of the oil, destroying the characteristics and delicate Estragon note.

According to Gildmeister and Hoffmann (1950), dried German herb material yields from 0.25 to 0.8% of oil and fresh material from 0.1 to 0.45% of oil. Chris (1923) reported a yield of 1.4% for fresh herb grown in Abadie (France). When dried for 24 h, the herb yielded only 1.04% of oil (calculated from the fresh herb weight). Plant material that has been cut during the flowering stage and reduced to clover dryness yielded 0.3–0.35% oil. According to Heeger (1940), Estragon yields from 1.35 to 2.40% of volatile oil. The so-called German Estragon contains more oil than the Russian varieties.

Physico-chemical Properties of Tarragon Essential Oil

Oil of Estragon is a light yellow to greenish liquid of peculiar, strongly aromatic odor, characteristic of the plant and reminiscent of Sweet Basil and Anise plants.

Table 7.5 summarizes the physico-chemical properties of Estragon oil as per the report of Gildenmeister and Hoffmann (1950).

Table 7.5 Physico-chemical properties of Tarragon essential oil

Property	Content
Specific gravity at 15 °C	0.900–0.945
Optical rotation at 20 °C	+20 to +90
Refractive index at 29 °C	1.504–1.516
Acid number	Up to 1
Ester number	1–9
Ester number after acetylation	15–22
Solubility	Soluble in 6–11 vol. of 80% alcohol, eventually with slight turbidity Soluble in 0.6–1.5 vol. and more of 90% alcohol

Methyl chavicol is the chief constituent of the volatile oil and phellandrene and ocimene are additional constituents.

To conclude, both yield and quality of the physico-chemical properties of essential oil are variable depending upon the basic quality of the raw material used, method, duration of distillation, etc.

Precautions needed in the storage of the Essential Oil

Oil of Estragon has to be stored in well-filled bottles, protected from light. When exposed to air, the chief constituent of the oil, methyl chavicol, is oxidized to an aldehyde, with accompanying increase in the specific gravity and refractive index of the oil. The solubility in alcohol improves to such an extent that an aged oil is soluble in 80% alcohol. Gildenmeister and Hoffmann (1950) reported that physico-chemical properties of an oil (1) after standing 6 months in an open bottle underwent considerable following changes (2), as summarized below:

Property	Situation 1	Situation 2
Specific gravity at 15 °C	0.9168	1.0475
Optical rotation	+3 degree 10	+3 degree 10
Refractive index at 20 °C	1.50847	1.52493
Solubility	Soluble in 1 and more vol. of 90% alcohol	Soluble in 1 and more vol. 80% alcohol

The aged oil (2) gave a positive reaction on treatment with bisulfite solution, which indicates that an aldehyde had been formed by air oxidation of the oil.

Uses

As A Food Flavorant

It is used to flavor vinegar ("Tarragon Vinegar"), pickles, prepared mustard, and to a limited extent to flavor soups, salads, meat dishes, certain cheeses, canned soups, table sauces, salad dressings and liqueur.

As a Pharmaceutical product

The aromatic leaves are credited with aperient, stomachic, stimulant and febrifuge properties.

In Perfumery

It is also employed in perfume manufacture.

References

Chris (1923) Perfums France 8: 23
Deel and Deel (1929) Bull Soc Chim 46:175
Gildmeister and Hoffmann (1950) Die Atherischen Ole, 3rd edn, vol III, p 1000
Heeger (1940) Pham Ind 7:372

Chapter 8
Thyme

Abstract At first, the chapter would discuss the description and distribution of the crop, followed by a discussion on area, production and trade. The chapter would then discuss crop management, post-harvest technology and processing technology. Discussion will also highlight the uses of Thyme products, especially its use in pharmaceutical products manufacture and as a food flavorant.

Keywords Description · Distribution · Crop management · Trade · Food flavorant

Botanical name: *Thymus vulgaris* Linn.

Syn. Thymus serpyllum Linn. (Wild Thyme)

Family: Labiatae

Description and Distribution

Thyme is one of the best-known European culinary herbs. The dried leaves and flowering to ps of *Thymus vulgaris* Linn. are called Thyme and those of *Thymus serpyllum* wild Thyme, which is found in the Western temperate Himalayas, from Kashmir to Kumaon hills, India, at an altitude of 1525 m above mean sea level. Thyme is grown in Europe, Austria and North Asia. It is now cultivated in France, Germany, Spain, Italy and other parts of Europe and, also, in England, North Africa, Canada and the USA. It is a common garden plant which survives for many years under good culture. The dried leaves are curled, are of brownish-green color, usually no longer than 6–7 and 2–3 mm broad and marketed in whole or in ground form. The flavor is aromatic, warm and pungent.

Wild Thyme is a low, hardy perennial, undershrub, 20–40 cm high, successfully introduced into the Nilgiris district of Tamil Nadu, in Southern India, or at higher elevations. It is used both for seasoning and for its volatile oil, ranging from 1.08–2.0% (average 1.0%), and in India, its cultivation on a commercial scale can be tried in the western temperate Himalaya from Kashmir to Kumaon hills between 1500 and 4000 m above mean sea level. The plant can be grown for ornamental edging along the paths in gardens and is also suitable for rockeries. Leaves are oblong-lanceolate, 10 mm × 3 mm, with orange-brown glandular dots; flowers are coriaceous, small,

K. P. Nair, *A Compendium of Unique and Rare Spices*,
https://doi.org/10.1007/978-3-031-20249-0_8

Table 8.1 World trade (exports and imports) of Thyme in 1976

Exporting country	Exports		Importing country	Imports	
	Quantity (tons)	Value (US$)		Quantity (tons)	Value (US$)
Spain	1309	865,000	USA	574.3	561,200
France	381	465,200	France	560.0	391,000
Lebanon	329	56,500	West Germany	275.0	288,000
Morocco	274	83,000	U.K	263.0	207,300
Turkey	196	256,200	The Netherlands	39.0	49,200
Syria	83	33,000	Belgium/Luxemburg	45.0	42,100
Total	2572	US$1.76 million		1756.3	US$1.54 million

lavender, purplish to almost white in color, attractive to honeybees, in pubescent verticillasters; nutlets are brown in color.

Area, Production and Export

Though official data on acreage and production of Thyme are unavailable, yet, quite useful information compiled by Greenhalgh (1979) clearly showed the importance of the popular Thyme herb in the world market (exports and imports) as summarized in Table 8.1. Thus, the total export of both cultivated and wild Thymes amounted to 2572 tons of dried Thyme leaves (crushed and ground), valued at about US$1.76 million. The most important producing and exporting countries of Thyme are Spain, France, Lebanon, Morocco and Italy. Likewise, the important Thyme importing countries are the USA, France, West Germany and the U.K.

An idea of the more recent trends (during 1983–85) in the world trade (import/export) of Thyme can be had from Table 8.2, giving both quantum and value of the herb imported/exported, along with the importing and exporting countries. A perusal of the data in the following table shows, in general, a mixed trend, varying with different countries.

During 1990–91, export of Thyme from India rose to over INR 3.9 million. The data are tabulated in Table 8.3.

Major Types of Thyme

French Thyme

Until a decade ago, France accounted for roughly 30–50% of the US Thyme supply, but, in recent years, it has fallen down to 10%. However, its product continues to command the highest price for its oil content, color and cleanliness. Its leaf particles average higher oil content than those from other sources and somewhat greener in color. Most French Thyme is still a product of wild plants, but some of its production recently comes from cultivation also.

The Provence region of France produces the *crème de la crème* of French Thyme; however, very little of true Provence Thyme is seen in the USA market. Provence

Table 8.2 Trends in world trade in Thyme (1983–85)

Major importing exporting countries and their currency	1985		1984		1983		Major exporting countries
	Qty	Value	Qty	Value	Qty	Value	
Belgium (BF)	96	9188	62	8850	72	6507	Germany Morocco France Spain
Denmark (DKr)	13	8	13	347	14	312	same as above
France (FF)	618	6102	835	6522	455	3490	Spain Morocco Albania
Germany (DM)	364	1628	443	2140	433	1989	Spain Poland Morocco
Italy (Lira)	19	22	39	42	25	29	–
The Netherlands (F)	81	416	96	385	98	441	Spain France Belgium Geneva
UK (Sterling Pound)[a]	202	211	177	203	118	120	Spain
Austria (S)	43	1848	47	1683	NA	NA	Spain Tukey Germany
Finland (Fink)[a]	99	1177	65	999	57	840	Geneva Turkey
Norway (Nkr)[a]	32	658	26	541	28	580	Not stated
Spain (Ptas)	32	3859	70	5946	39	4565	Morocco Uganda
Switzerland (SWF)[a]	89	415	103	527	92	436	Turkey Spain France Ethiopia
USA (US$)[b]	10,451	19,419	12,174	19,160	12,215	19,860	Not Stated
Brazil (US$)[a]	5	11	3	5	14	16	Turkey Portugal France Argentina
Morocco (DH)[a]	–	1	–	–	–	2	Not stated
Jordan (JD)	270	43	379	49	297	43	Syria Lebanon Turkey
Saudi Arabia (SRTS)	392	1561	468	2214	390	2033	Lebanon Jordan India Syria

(continued)

Table 8.2 (continued)

Major importing exporting countries and their currency	1985		1984		1983		Major exporting countries
	Qty	Value	Qty	Value	Qty	Value	
Hong Kong ($HK)[a]	25,059	1,52,510	20,720	156,244	15,599	112,876	China USA Vietnam Korea

Notes [a] Thyme and Bay together
[b] Thyme, other herbs and minor spices
Qty—Quantity in tons (000)
Figures in parentheses indicate the respective currency of the country involved.
Source International Trade Center, Geneva (ITC), UNCTAD/GATT (1986): Imports of Spices into Selected Markets during 1981–85. DOC. Presented at the 1st Meeting ISG, New Delhi, India, November 1986. pp.1–94

Table 8.3 Export of Thyme from India (1990–91)

Destination country	Quantity exported (kg)	Total value(INR)
Australia	200	3979
Korean Republic	200	4384
Malaysia	28	3000
Oman	76	1600
Singapore	100	1550
United Arab Emirates	4000	43,800
USA	5384	71,822

Source APEDA, New Delhi, India

Thyme has a very high oil content, and consequently a much stronger flavor, and the flavor is also known as particularly pleasant. It is mountain-grown and comes from wild plants.

Spanish Thyme

Spain is by far the largest exporter of Thyme today. It harvests three seasonal types (all *Thymus vulgaris*): (1) Winter (February–March), (2) Gray (April–May) and (3) Red (May–June). The names, however, are not descriptive, all three types look very similar. The Red type has the highest oil content and is used primarily for distillation. The Gray is favored for the spice trade, being a product which is handled and cleaned carefully and also has a good oil content. The Winter type is the least expensive product-lowest in oil content and less attractive. It should be noted that these distinctions are made primarily within the spice trade. The food company buyer is more apt to find such terms as Spanish Fancy and *FAQ* (*Fair Average Quality*—a term referring mostly to the cleanliness).

American/California Thyme

Thyme is also being grown commercially in California. Most of it is *Thymus vulgaris* of the type known as French Thyme. Additionally, a small amount of *Thymus citriodorus* (Lemon Thyme) is produced in the Golden state—mostly for use in herbal teas, to which it brings a strong lemon aroma. All the Californian Thyme is cultivated, yielding about three crops per year. As with other domestic herbs, the American Thyme is machine-dried, cleaned and selected for its rich color as well as flavor.

Crop Management

Climate and Soil

Thyme prefers a mild climate and a mallow upland soil, but it can grow in poor dry soil, as well. The plant grows best in hills. In the plains, good care is needed and the Thyme plant should, preferably, be grown in pots.

Propagation

It may be propagated from seed or cuttings. The seeds are drilled into the holes in rows about 90 cm apart. The seedling may be thinned out to 30–45 cm apart. Mulching in the winter prevents damage by frost, in particular, in the hills. The crop requires inter-culturing and weeding throughout the growing season. The leaves and flowers, which are used for culinary and medicinal purposes, are ready for harvest in about 5 months after sowing. They are plucked, dried in shade and stored in air-tight containers to prevent the loss of flavor. The dried shoots may also be powdered and packed. No statistics of yield in India are available on Thyme, but it is reported that in other countries where it is grown, a yield of 1100–2200 kg of dry herb/ha can be expected under favorable conditions; the yield of oil per hectare is not likely to exceed 22 kg. The yield is comparatively low during the first year after sowing. Since the plants have a tendency to become woody, their replanting becomes necessary after 3–4 years following sowing.

Post-harvest Technology

Commercial Quality

The dried leaves and floral tops constituting the Thyme plant of commerce are also known as "*Thymi Herba*", in pharmaceutical parlance. The drug occurs chiefly as leaves, mixed with a small amount of segments of the floral tops and flowers. It should contain not more than 3% of stems over 1 mm in diameter and 2% of other foreign organic matter; it yields 4% acid-insoluble ash. Thyme possesses agreeable odor and an aromatic and warming taste. Owing to the presence of Thymol, it shows germicidal properties and is effective against a variety of pathogenic bacteria. Extracts of the shoots have shown antibacterial action against *Micrococcus pyogenes* var. *aureus* and *Escherichia coli*, and this activity is maximum in the extracts from the flowering plants.

Table 8.4 Physico-chemical properties of Thyme

Property	Content
Moisture	7.1%
Protein	6.8%
Fat	4.6%
Crude fiber	24.3%
Carbohydrates	44.0%
Mineral matter	13.2%
Calcium	2.1%
Phosphorus	0.20%
Iron	0.14 mg/100 g
Sodium	0.08%
Potassium	0.9%
Vitamin A	175 IU (International Units/100 g)
Vitamin B_1	51 mg/100 g
Vitamin B_2	0.4 mg/100 g
Vitamin C (ascorbic acid)	12.0 mg/100 g
Niacin	4.9 mg/100 g
Calorific value	340 cal/100 g

Composition: Table 8.4 summarizes the physico-chemical properties of Thyme.

Handling and Storage of Thyme

Thyme and its extractives should be stored under cool, dark, dry conditions to protect them against fragrance and flavor losses. As with other any highly aromatic herbs, already discussed in this book in earlier chapters, Thyme leaves should be further protected against flavor losses by making sure that the container in which they are stored is completely air-dry and tightly closed after every use. Under good storage conditions, the Thyme leaf will retain its flavor, qualities and strength long enough to meet all normal food manufacturing stipulations.

Processed Products

Thyme Extractives—Volatile Oil and Oleoresins

Both essential oil and oleoresin extractives are prepared from the flowering tops of the Thyme plant. As with other aromatic herbs, the volatile oil is the most important extractive. The Spanish Red Thyme is the prime source of the oil, and it is marketed as a Red Thyme Oil. There is also a White Thyme Oil, but this is a redistillation product of the Red Thyme Oil rather than the product of another plant. It tends to be smoother in flavor than the Red Thyme Oil, and is, pleasingly aromatic. Thyme oleoresin is discussed subsequently.

Crushed Thyme, on steam distillation, yields about 2.5% volatile oil, of which is a colorless yellowish-reddish liquid with a pleasant odor, characteristic of the herb and with a sharp taste. The volatile oil is found to have the following properties:

1. Specific gravity at 15 °C 0.905–0.930
2. Optical rotation at 20 °C: 1.48
3. Solubility: 1 part in 2 parts of 80% alcohol.

Thymol

The chief constituent of the oil is Thymol (about 45%). In the pure state, Thymol consists of colorless, translucent crystals which possess an aromatic Thyme-like odor and a pungent taste. Thymol is now produced synthetically.

Thymol is in great demand in India as can be judged by the import figure. During 1971–72, India imported 34,204 kg of Thymol valued at INR 11,00,000 (INR 1.1 million). An increase in Thyme acreage should be seriously considered in India.

The wild Thyme yields 0.3% essential oil (as compared to 2.5% in cultivated Thyme), containing phenol and other compounds.

Oleoresin of Thyme contains both volatile and nonvolatile components. As extracted, it is a very viscous material which is normally standardized with a food-grade solvent before use. The oleoresin is often added to soluble carriers, either in liquid form or dry form, for seasoning applications in food preparation.

Uses

As A Food Flavorant

The herb is used to season tomato soups, "clam-chowder" and juices, fish and meat dishes (in particular poultry dressing). It is also used in liver sausage, pork sausage, head cheese, cottage and cream cheese and bockwurst.

As A Pharmaceutical Product

The oil of Thyme is well recognized for its medicinal properties. It is employed in preparations for use in the treatment of bronchitis and whooping cough. The oil of wild Thyme is also used to relieve tooth ache.

In Punjab State, India, the herb of wild Thyme is used to correct weak vision, complaints of stomach and liver, suppression of urine passage and also in menstruation.

In Europe, it is considered as an antispasmodic and carminative.

An infusion of Thyme herb is administered to remedy convulsive coughs, whooping coughs, catarrh and sore throat. It is good to remedy nervous or hysterical headaches and also to control flatulence.

The herb has a sharp pleasant taste, is emmenagogue, alexiteric and anthelmintic, and also effective to remedy liver complaints, pain in spleen, liver and chest. It is also useful in remedying asthma and bronchitis. It thins phlegm and blood. The leaves are a laxative, stomachic, tonic, good for the kidney and eyes, and are blood purifier.

It soothes the throat and cures cough; strengthens lungs, prompts good sleep and rids of nightmares; improves digestion, cures hangovers and helps the dull sightedness; warms the heart, soothes the liver and takes away hardness of the spleen. It is an excellent remedy for shortness of breath; "provokes the terms"; takes away hot swelling, eases the gout and cures pains in the loins and hips. It also purges phlegm.

The seeds are given as a vermifuge. Its infusion is also used to remedy skin eruptions/diseases.

Reference

Greenhalgh P (1979) Markets for culinary herbs. Trop. Prod Institute, London Special Market Report No. G121(including Thyme)

Chapter 9
Calamus or Sweet Flag

Abstract At first, the chapter would discuss the description and distribution of the crop. Details on area, production and trade would follow. This would be followed by a discussion on crop management and post-harvest technology. Discussion on processed products would lay stress on essential oil of Calamus.

Keywords Description · Distribution · Crop management · Processed products · Essential oil

Botanical Name: *Acorus calamus* Linn

Family: Araceae

Description and Distribution

Sweet Flag or Calamus is reported to be common in Kashmir and Kumaon ranges of the Himalayan forests. The rhizomes are collected from the Himalaya forests and brought to Kathgodam, from where they are dispatched to Mumbai, Kolkata and Kannauj.

It is a semi-aquatic perennial herb with a creeping and much-branched aromatic rhizome. The rhizome is cylindrical or slightly compressed, about 19–25 mm in diameter and about 10 cm long. Externally, it is light brown or pinkish in color, and white and spongy within. It is thoroughly washed in water and freed from dirt, and rootlets are removed and then carefully dried at a low temperature. It deteriorates with age and is subject to attack by worms. It is this dried rhizome, which constitutes Calamus or Sweet Flag, which is used as a condiment and as a pharmaceutical product. It has a reputation as a healing herb for its use in medicine.

It is found wild as a water weed in shallow margins of creeks, lakes, pools and some rivers, throughout India and Sri Lanka, ascending to 1830 m above mean sea level in the Himalayas. The plant thrives in marshy places and in most situations, like the edges of lakes and banks of streams. It is cultivated in one district, Koratagere, in Karnataka State, India. Curiously enough, the cultivation of Sweet Flag has not been extended to any other part of the State. It is also cultivated in Europe and USA, more for medicinal purposes.

K. P. Nair, *A Compendium of Unique and Rare Spices*,
https://doi.org/10.1007/978-3-031-20249-0_9

Area, Production and International Trade

No reliable information is available on any of the above aspects in India, except that it is traded both for medicinal and condiment purposes.

Crop Management

The Calamus plant is grown very much the same way as rice in the fields, which are puddled, and kept irrigated like rice in the field. This crop occupies the field for the whole year.

Climate and Soil Requirements

It thrives best in marshy and moist places under variable climate. It is regularly cultivated in Koratagere district of Karnataka State, where the plant is grown in clayey loam and in light alluvial soils of the river banks.

Preparation of Land

The field is irrigated sufficiently to soften for plowing and after being plowed twice, it is watered heavily and plowed in puddle and left, as such, for a few days and plowed in puddle, once again. It is then leveled by a leveling board over the puddled soil. The field is then heavily manured with the leaves of "honge" (*Pongamia glabra*), which are trampled into the puddle. In this condition, the field is ready for planting.

Planting Method

The planting material is made up of the growing ends or tops of the previous crop of Calamus, much as in the case of sugarcane crop. At the harvest time, the mature portion of the rhizome is cut off, which actually forms the marketable crop. The remaining tender portion, formed by the growing end, is reserved for planting. These tops can be planted at once or can be kept covered up with dry leaves or straw on the ground for about a week or ten days without being damaged. If they have to be kept for a longer period, then they are kept in open pit, well moistened with water. At the planting time, these tops are slightly trimmed by cutting off and shortening leaves at the ends and are planted in the soft mud by pressing about 5 cm of the root into the soft mud, the leaf growing end being well above the ground. The planting is so done that this end slopes and points away from the direction in which water enters and flows into the bed, with the belief that it helps in making the rhizomes grow longer.

The tops are planted in rows, 30 cm apart and, at a distance of 30 cm from each other in the rows in the "scissor style", that is to say, that in the plants in the second row come in between the plants of the first row and not opposite to them. Thus, plants of only alternate layers are exactly opposite to each other.

Planting Time

The planting season is March–April, but the crop may be planted in any part of the year, keeping in view the fact that the rhizomes have to be dried in the sun after they are harvested and one has to ensure that the harvest may not coincide with the rainy season.

Irrigation

The field is kept regularly irrigated and the level of the water maintained as in the case of rice, that is, at least to a depth of 5 cm in the beginning of the crop and increased to 10 cm as the plants grow in height.

Weeding

The field is also weeded several times carefully, and, every time, the growing plants are somewhat pressed down along their length into the soil, with only the leaves and a little of the stem growing erect. Weeding is done once a month, and usually, 8 weedings are done during the entire period of crop growth.

Manuring

In the fourth month, the field is manured with cattle manure or FYM (Farm Yard Manure), at the rate of 4 cartloads per acre. The organic manure is broadcast before weeding.

Harvest

In about one year since planting, the tops begin to get yellow and are ready for harvest. Irrigation is stopped and the field is allowed to dry, somewhat, retaining just enough moisture to make the digging of the rhizomes easy. The digging has to be done quite deep, which is quite a laborious procedure and, hence is cost intensive, as the digging is done manually. This is also because the rhizomes send down the roots quite deep and profusely, as deep as 60 cm. After the plants are completely dug out, the tops are cut off and retained as seeds for further planting in the following season.

The rhizomes are about 30–45 cm in length and are cut into short lengths of about 5–8 cm each, and all the fibrous root growth on these pieces is cut away. The cut pieces are then washed to remove adhering soil and put out in the shade for drying.

The recent systematic investigations conducted on the effect of planting materials and time of harvest on the growth, yield of rhizomes and essential oil content in the rhizomes of Calamus (*Acorus calamus* Linn.) by Philip et al. (1989) in the State of Kerala, India, have revealed that the growth of the crop is significantly influenced by the type of the planting materials used. The planting materials and time of harvest had significant influence on the drying percentage and essential oil content of the rhizomes, whereas the type of planting materials had no effect in this regard. The drying percentage of the rhizomes increased with the increase in the maturity of the crop, whereas the essential oil content of the rhizomes showed an inverse relationship with the maturity of the crop. The interaction effects of planting materials and dates of harvest on the yields of green rhizomes, dry rhizomes, drying percentage and the essential oil content of the rhizomes were found to be nonsignificant. The use of rhizome bits of 6 cm length with the growing tops and leaves as planting material and harvest of the crop 10 months after planting were found to be optimum for realizing maximum yield of the rhizomes and their essential oil content. The presence of growing tops and leaves in rhizome bits used for planting had significant influence on growth and yield of the crop. Hence, the retention of the growing tops and leaves

Table 9.1 Effect of planting materials on growth of *Acorus calamus* Linn

Treatments	Plant height (cm)	Tiller number per plant	Leaf number per tiller	Leaf number per plant	Leaf area (cm^2)
3 cm rhizome bits with growing tips	9.6	6.2	5.5	33.5	73.7
6 cm rhizome bits with growing tips	11.8	7.3	5.6	40.4	75.1
3 cm rhizome tips without growing tips 8.4	8.4	4.4	5.5	22.9	50.9
6 cm rhizome bits without growing tips	7.5	5.1	6.0	29.8	56.7
Critical difference (P = 0.05)	1.3	1.0	N.S	5.8	8.9

(Mean values)
Notes N.S. Not significant
Source Philip et al (1989)

on the rhizome bits, used for planting, is essential for getting better growth and higher yield of rhizomes. The results are summarized in Tables 9.1 and 9.2.

Yield

An acre of the crop yields about 1.5 tons of dry marketable rhizomes. A very good crop may yield up to 3 tons. The price of the product fluctuates. Nevertheless, it is claimed to be quite remunerative to grow Sweet Flag (Philip 1989).

Plant Protection

The crop is practically free from pests and diseases, apart from stray attacks of a "leaf spot" disease, which does not adversely affect the growth or the yield of the Sweet Flag crop, in any manner. An aphid-transmitted mosaic streak disease of wheat is also reported to affect the crop.

Post-harvest Technology

The green rhizomes are cut into pieces of 5–8 cm long, and all the fibrous roots/rootlets are removed. The cut pieces are washed in water thoroughly and dried in the sun. The dried material is put into rough gunny bags and rubbed to remove leafy scales. The dried rhizomes constitute the spice as well as the drug—the Calamus of the trade. They are sometimes substituted for the rhizomes of *Alpinia galanga* (Linn.) Wild (Bach) and *Aconitum* spp. (Bish). The mixture is sold in the Indian market, as such.

Table 9.2 Effect of planting materials and time of harvest on yield of rhizome and essential oil in rhizome of *Acorus calamus* Linn (Mean values)

Treatments	Yield of green rhizomes (g)	Yield of dry rhizomes per plot	Drying percentage	Green rhizome yield/ha (kg)	Dry rhizome yield/ha (kg)	Essential oil content (%)	Yield of essential oil/ha (kg)
I. Time of harvest							
Harvest 8 months after planting	116.8	59.7	51.2	935	478	1.55	7.4
Harvest 9 months after planting	134.0	71.8	53.4	1072	574	1.38	7.9
Harvest 10 months after planting	136.8	79.3	58.0	1094	634	1.33	8.4
harvest 12 months after planting	114.3	69.5	60.2	914	596	1.30	7.2
Critical difference (P = 0.05)	16.1	8.4	3.0				0.14
II. Planting materials							
3 cm long rhizome bits with growing tips	112.0	62.7	56.4	896	502	1.45	7.3
6 cm long rhizome bits with growing tips	172.5	93.9	54.7	1360	751	1.391	0.4
3 cm long rhizome bits without growing tips	92.8	51.6	55.8	742	413	1.37	5.7
6 cm long rhizome bits without growing tips	128.5	72.0	56.0	1028	576	1.35	7.8

(continued)

Table 9.2 (continued)

Treatments	Yield of green rhizomes (g)	Yield of dry rhizomes per plot	Drying percentage	Green rhizome yield/ha (kg)	Dry rhizome yield/ha (kg)	Essential oil content (%)	Yield of essential oil/ha (kg)
Critical difference ($P = 0.05$)	16.8	9.1	N.S.			N.S.	

N.S. Nonsignificant
Source Philip et al. (1989)

Table 9.3 Composition of Calamus rhizome

Property	Content (%)
Moisture	66–71
Starch	7.5–13.1
Ash	1.24–1.3
Dry matter	26–41

Adulteration

The powdered drug is reported to be adulterated with siliceous earth, ground root of "Marsh Mallow" (*Althaea officinalis* Linn.) and cereal flours.

Composition

Published information on the composition of Calamus is rather sparse, though sufficient data are available on its oil content, which will be discussed subsequently in this chapter. According to a German report, considerable variation is noticed in the composition of the Calamus rhizome, which is as follows (Table 9.3).

Processed Products

Volatile Oil

The dry rhizomes, on steam distillation, yield 1.5–3.5% of a yellow aromatic volatile oil. It has a mellow odor, resembling that of *patchouli* (*Pogostemon cablin*, family Lamiaceae, perennial mint). Prior to charging the steam-distillation still, the rhizomes are reduced to the desired size in a suitable disintegrator. The yield is about 4.6%.

Indian Calamus oil has the following properties (Table 9.4).

Indian oil has a much higher "Asarone" content (82%) than the commercial oils (7–7.5%). The characteristic odor of the rhizome and oil is ascribed to an unidentified constituent of boiling point of 125–135 °C. Table 9.5 summarizes the variations in the composition of Calamus oil.

Table 9.4 Properties of Indian Calamus oil

Property	Content
Specific gravity at 15 °C	0.958–0.970
Refractive index at 30 °C	1.500–1.508
Optical rotation	+ 9° to 35°
Acid value	0.3–2.4

Table 9.5 Variations in the composition of Calamus oil

Source yield (%)	Specific gravity at 15 °C	Refractive index	Optical rotation	Acid value	Ester value	Saponification
Bengaluru (a) (India)	2.8	1.076	1.5461 at 30°	−1° 50′	2.4	−4.1
Jammu (b) (India)	3.1	1.056	1.5540	+ 2 (at 15°	0.14	1.4
Kashmir (b) (India)	1.4	0.971	1.5306	+ 14°	2.24	12.7
Europe (c)	0.94–2.2	0.960–0.974	1.5045–1.5070 +	+ 15° to 18°	40 up to 3.7′	6.8–7.5
America (d)	3.3	0.950–0.974	1.5013–1.5069	13° to 48° to 15°	(at 20 °C)	8.4–10.7
Japan (a)	4.63	0.975–0.985	1.5051–1.5090	2° 2°	80′ to + 13′40′	
Russia (d)	3.58–7.80	0.936–0.955	1.5029–1.5130	1.6–3.9	7.6–13.9	5.8–12.2
Yugoslavia (e)	1.3–4.1	0.960–0.971 (at 20 °C)	1.50 (at 20 °C)	−13° to 1–2	−27 °C	7–15

Source a. Kelkar and Rao. J. Indian Inst. Sci. 1934, 17A, 25
b. Chaudhury et al. Indian J. Pharm. 1957, 19,183
c. Guenther, VI, 109
d. Chem. Abstr. 1972, 76.138157
e. Tucakov. Perfum essent Oil. Rec.1958. 49,491

Uses

The medicinal properties of the Calamus plant have been known since the earliest time of human civilization. There are references to it in the early ***Sanskrit*** (ancient Indian language) and Chinese literature. Its medicinal action is attributed to its essential oil content. The roots also contain a glucoside called "acorin". The oil is reported to cure gastritis and is supposed to promote digestion also. It is also administered to animals as a stomachic. The essential oil is used to cure stomach and skin diseases, as a vermifuge, and, an antiseptic, as well.

The dried rhizome is a common *bazar* (Hindi name for an Indian market) medicine and is generally used in the form of infusion. It is aromatic, bitter, toxic and a carminative. It possesses emetic and antispasmodic properties. It produces beneficial results in cases of dyspepsia and chronic diarrhea. In large doses, it acts as an emetic. It is also used in the treatment of insecticidal worms, the bites of poisonous reptiles, foul breath, throat infections, etc.

Bakhru (1972) has reviewed all of its important medicinal virtues and its usefulness in severe health disorders and diseases like stomach disorders, mouth disorders, cold and whooping cough, diarrhea and dysentery, mental diseases, intestinal worms, breaking tobacco addiction and in an array of other uses.

In Perfumery

The greatest use of Calamus or Sweet Flag at present appears to be in the perfumery industry, as a component of the finest perfumes. It also enters into the composition of face powders.

As An Insecticide

The powdered rhizome also possesses insecticidal properties and is useful against bed-bug, moths, lice, etc. It also forms the main ingredient of many insect powders, used to protect clothes, and is kept in cupboards where clothes are kept, to protect them from insect damage.

In Foods, Beverages and Liquors

The rhizome and its oil are used in the manufacture of liquor, essences and bitters.

Reference

Philip J, Nair GS, Premalatha, Sudhadevi PK (1989) Effect of planting materials and time of harvest on yield and essential oil content of rhizomes in *Acorus calamus* Linn. Indian Cocoa Arecanut and Spices J 13(2):63–65

Chapter 10
Horse-Radish, Long Pepper and Galangal

Abstract *Horse-Radish* At first, the chapter would discuss the description and distribution of the crop, to be followed by discussion on crop management and post-harvest technology. Discussion would also be on adulteration of the crop and its processed products, with special emphasis on the use of the crop as a condiment. *Long Pepper or Pipli* To begin with, the chapter would discuss the description and distribution of Pipli. This would be followed by a discussion on crop management, area, production, trade and post-harvest technology and processed products. Discussion on use of Pipli in pharmaceutical products manufacture and as a food flavorant will also be included. *Galangal* At first, the chapter would discuss description and distribution of the crop. A discussion on the economic importance of galangal would follow. Further discussion will be on crop management, post-harvest technology, etc. Detailed discussion on the volatile oil of galangal would also be included in the chapter.

Keywords for Horse-Radish Description · Distribution · Post-harvest technology · Adulteration · Condiment
Keywords for Long Pepper or Pipli Description · Distribution · Trade · Post harvest technology · Food flavorant
Keywords for Galangal Description · Distribution · Crop management · Post harvest technology · Volatile oil

Horse-Radish
Botanical name: *Cochlearia armoracia* Linn

Family: Brassicaceae

Description and Distribution

A near relative of turnip, cabbage and mustard. It is one of the oldest condiments and is a well-known, large-leaved, hardy perennial in many long-established gardens. It is the thick, white, fleshy, tasty root of the Horse-Radish, which is highly prized as an appetizing condiment which goes well with many food items.

A native of the marshy districts of Eastern Europe, it is grown in the USA and to some extent in the gardens, both in the Northern India and in the hill stations of the Southern India. In the USA, about 7 million kilograms of Horse-Radish are processed

K. P. Nair, *A Compendium of Unique and Rare Spices*,
https://doi.org/10.1007/978-3-031-20249-0_10

Table 10.1 Composition of Horse-Radish

Property	Content (%)
Moisture	73.4
Protein	3.2
Fat	0.2
Ash	1.8
Total carbohydrates	21.4
Crude fiber	2.4

The root is a rich source of vitamin C (ascorbic acid), and the fresh material contains, on an average, 302 mg of ascorbic acid/100 g

annually for consumption as food. Cultivated over 1500 acres in the USA, its varieties are classified into two types, namely Common and Bohemian. The Common type has, broad, crinkled leaves and produces a root of high quality. The Bohemian type has narrow, smooth leaves and is disease-resistant, but yields poorer quality roots. Horse-Radish plant is highly sterile and hence is propagated asexually by the root. The tap root is tuberous and cylindrical, 30 cm long and about 18 mm in diameter, possesses an acrid, pungent taste and, when scraped or bruised, emits a characteristic pungent odor (Guenther 1973).

The root contains a pungent, acrid and vesicating volatile oil. The pungency is due to the presence of "Sinigrin"(myronate potassium), a sulfur-containing glucoside, which, in the presence of water and under the influence of an enzyme myrosinase, yields allyl mustard oil or allyl isothiocyanate, as by mustard seeds (Willstater and Stoll 1918; Stoll and Seeback 1948).

Area, Production and Export

Horse-Radish, being a garden crop, so far, there is no recoded data on any of the aspects of area, production and export from India.

Crop Management

As the Horse-Radish is not being commercially cultivated, the precise information on the cultural aspects is not readily available.

Post-harvest Processing Technology

Volatile Oil

Distillation of the titrated roots gives about 0.05–0.2% volatile oil, which is not commercially produced.

Composition of Horse-Radish

Table 10.1 details the composition of Horse-Radish.

Adulteration

Prepared Horse-Radish is sometimes adulterated by admixture with turnips and parsnips, followed by fortification with allyl isothiocyanate (synthetic mustard oil), to produce desired pungency. It is rather difficult to detect this type of adulteration precisely. However, it may be stated that Horse-Radish contains from 0.22 to 0.86% of the natural allyl isothiocyanate on moisture-free basis, while parsnips and turnips contain practically none (less than 0.08%). Likewise, Horse-Radish also contains considerable starch content (23%), while turnip contains less than 10%, and parsnip contains from 11 to 18%. Besides, the infra-red spectrum of volatile oil of allyl isothiocyanate differs distinctly from the spectrum of volatile oil of parsnip. With the help of the above criteria, it may be possible to pinpoint the nature and the extent of the type of adulteration.

Uses

As A Condiment

Horse-Radish is used as an appetizer condiment. It is credited with digestive and antiscorbutic properties because of its high vitamin C content. It is a highly prized condiment, especially when used with oysters and meats. Besides, leaves and roots of the plant are used as a food tm in Germany.

A freshly grated Horse-Radish root, when mixed with vinegar and salt, is much appreciated as an appetizing and pungent condiment which enhances the flavor of the boiled or roasted beef. When mixed with ketchup, the grated root imparts a refreshing taste to seafood, especially shrimp cocktail and oysters.

Horse-Radish Cream, Sauce or Relish

Horse-Radish is used for the preparation of "Horse-Radish Cream", "Sauce" or "Relish", which is rather a salad-cream and contains at least 25% shredded Horse-Radish (freshly shredded) or equivalent of dehydrated root, essential oil, cream containing at least 25% edible vegetable oil, 7.5% of nonfat milk solids (NFMS). It includes egg yolk, milk, vinegar, salt and mustard also.

In Pharmaceutical Use

It is a stimulant, diaphoretic, diuretic and digestive. It is also used as an irritant in Lumbago and similar painful afflictions. Crushed Horse-Radish has an inhibitory effect on the growth of microorganisms, this effect being attributed to allyl isothiocyanate.

Long Pepper or Pipli

Botanical name: *Piper longum* Linn

Family: Piperaceae

Description and Distribution

Long Pepper or *Pipli* is the dried fruit of *Piper longum*, which is a slender, aromatic climber with perennial woody roots, occurring in the hotter parts of India, from the

Central Himalayas to Assam, Khasi and Mikir Hills and Lower Hills of Bengal and evergreen forests of the Western Ghats from Konkan to Travancore in Kerala State, India; it has also been recorded from Nicobar Islands. The plant has creeping jointed stems; young shoots are downy; leaves are 5–9 cm long, 3.5 mm wide, ovate, cordate, with broad rounded lobes at the base; female spikes are cylindrical, pedunculate; male spikes are larger and slender; female spikes are 1.3–2.5 cm long and 4.5 mm in diameter; fruits are ovoid, yellowish orange, minute, droop and are sunk in fleshy spike. The female spikes are sharper, sessile and blunt at the tips and erect and are red when ripe; flowers are numerous and packed.

Indian Long Pepper is mostly derived from the wild plants, the main sources of supply being Assam State in India, followed by West Bengal, Nepal and Uttar Pradesh. Small quantities are also available from evergreen forests of Kerala State, West Bengal and certain parts of Andhra Pradesh. It is reported to be cultivated at lower elevations in the Annamalai hills in Tamil Nadu, India, and in parts of Assam, in particular, the heaviest (in India) rainfall region, Cherrapunji. It is cultivated in Meghalaya also; India produced about 200 tons in 1990–91.

The Long Pepper, as sold in India, appears to be derived from 2 or 3 species, including one which is Indonesian. Indian Long Pepper is a product of either *Piper longum* or *Piper peepuloides*, while the Indonesian or Java Long Pepper imported from Malaysia is from *Piper retrofractum*. The fruits of these species are used for the same purpose, though they do vary in their comparative effectiveness.

In some hilly areas of Vishakapatnam in Andhra Pradesh, India, Long Pepper is grown for its roots and stems used in medicinal purposes. The roots constitute the important drug known as "*Piplamul*", used in ancient Indian Ayurvedic and Unani medicines. Obviously, the cultivation of Indian Long Pepper for "*Piplamul*" falls outside the scope of this chapter.

Area, Production and Trade

No official statistics are, yet, available on the extent of area under Indian Long Pepper or its total production in India. However, the Deputy Director of Spices Board in Guwahati, North East India, has reported that Indian Long Pepper is now produced in Meghalaya. Official statistics of production of this spice are yet unavailable. However, trade estimates indicate that 200 tons on Long Pepper are produced in Meghalaya annually, alone, though its production still remains confined in an unorganized manner. It has been reported that Indian Long Pepper produced in Meghalaya is of the best quality, sought by the Indian traders of the spice in Mumbai and Chennai, in India. The present price of this rare spice is reported to be INR 48/kg in the Shillong market in Assam District. Of course, India imports a large quantity of Long Pepper from Malaysia, Java and Singapore. Thus, during 1981–82, the imports amounted to 220,312 kg valued at INR 610,000. A small quantity of Indian Long Pepper has been exported from India in some years to Afghanistan, Sri Lanka and Pakistan.

Probably, the imports of Indian Long Pepper are mostly of the Java type from *Piper retrofractum*, which is a cheap variety, whereas exports of Indian Long Pepper from India are of *Piper longum* and *Piper peepuloides*.

Crop Management

Soil and Climate

Indian Long Pepper is cultivated on a large scale in limestone soil, 450–600 m below the Cherrapunji region, which receives the highest rainfall in India, from the end of March to middle of September, and this leads to a very high relative humidity in the region.

Propagation

Indian Long Pepper is propagated mainly by "Layering", a technique explained in the book by this author, titled "Minor Spices and Condiments—Global Economic Potential" (Nair 2022), of the selected, mature, healthy branches or by "suckers", planted at the beginning of the rainy season. But, these methods are time consuming. Recently, a method of large-scale rapid multiplication of Indian Long Pepper has been developed by Sarasan et al. (1993) in Kerala State, India. This has been achieved by plant regeneration in Long Pepper through indirect adventitious shoot development.

Manuring

The vines are well manured with Farm Yard Manure (FYM) or cattle manure.

Harvest

The vine starts to bear spikes after 4 years from the planting date. The spikes are harvested in January each year, while they are still green and unripe, but fully mature, as they are most pungent at this stage.

Yield

The yield of dried spikes increases from about 560 kg/ha in the first year to about 1680 kg/ha in the third year. Thereafter, the vines become less productive and should be replaced by fresh planting.

Plant Protection

The occurrence of leaf spot caused by the fungus *Cercospora* sp. in Indian Long Pepper (*Piper longum* Linn.) has been reported in Maharashtra State in India. No other published reports are available on pests and diseases attacking the vines, so far.

Post-harvest Technology

Drying and Packaging

The harvested green spikes are dried in the sun when they turn gray. They are packed in gunny bags. They have a pepper-like pungent taste and produce salivation and numbness of the mouth, when ingested.

Composition

Recent research on the fruit of *Piper longum* has shown the presence of alkaloids, piperine (4–5) and piplartine (melting point 124–125 °C), and two new liquid alkaloids, one of which is designated as alkaloid A. This is closely related to pellitorine

Table 10.2 Characteristics of long pepper essential oil

Property	Content
Specific gravity at 15 °C	0.8484
Refractive index	1.4769
Optical rotation	40.1°
Acid value	7.2
Saponification value	8.9
Saponification value after acetylation	12.8

Soluble in 20 vol of 50% alcohol

producing marked salivation, numbness of the mouth and a tingling sensation on the mucous membrane of the mouth.

Processed Products

Volatile Oil

Dried fruit of Indian Long Pepper on steam distillation gives 0.7% of an essential oil with spicy odor, resembling that of pepper or ginger oils, and having the following characteristics as detailed in Table 10.2.

Uses

As A Food Flavorant

The fruits of Indian Long Pepper are used as a spice, and they also are used in pickles and preserves. They have a pungent pepper-like taste and produce salivation and numbness of the mouth when eaten or munched.

In Chhota Nagpur, the root is used to ferment rice beer. In Andaman Islands, the leaves are chewed like betel leaves.

As a Pharmaceutical Product

The fruits as well as the roots of Indian Long Pepper are credited with numerous medicinal values and may be used to cure diseases of the respiratory tract, such as cough, bronchitis, asthma. It is also used as counter-irritants and as analogists when applied locally to cure muscular spasms or pains and inflammation; also it can be used as snuff in coma and drowsiness and internally as a carminative; as a sedative in insomnia and epilepsy; as a general tonic and hematinic. It is used also as a cholagogue in obstruction of the bile duct and gall bladder. It is an emmenagogue and abortifacient and also used for miscellaneous purposes such as anthelmintic and as a cure for dysentery and leprosy.

"Alkaloid A" naturally present in Indian Long Pepper exhibited significant in vitro antitubercular activity against *Mycobacterium tuberculosis* H-37 Rv. strain and inhibited the growth of the bacillus in a 20 g/ml concentration.

Alcoholic extracts of the dry fruits and aqueous extracts of the leaves showed activity against *Micrococcus phygenes* var. *aureus* and *Escherichia coli*. Ether extract of the fruits showed larvicidal properties.

Besides fruits, the roots and thicker parts of the stem are cut and dried and used as an important drug ("Piplamul") in the ancient Indian systems of medicine, namely *Ayurveda* and *Unani.*

Galangal

Botanical name: *Alpinia galanga* Linn. Willd

Family: Zingiberaceae

Description and Distribution

Galangal is the dried rhizome or root of a plant, *Alpinia galanga* Linn. Willd, which grows mainly in the Eastern Himalayas and Southwestern India. The plant is 1.8–2.1 m high and bears perennial rhizomes which are deep, orange brown in color, aromatic, pungent and bitter. The fruits are about 13 mm long, constricted in the middle and contains 3–6 seeds. The seeds are slightly pungent, with an aroma similar to that of the rhizome. Cut pieces of the rhizomes of this species are known as "Greater Galangal". The rhizome of "Lesser Galangal" is smaller and reddish brown in color and has a stronger odor and taste. According to Guenther (1973), Galangal root is also known under the names of "Smaller Galangal" and "Radix galangal minoris".

The earliest reference to this spice is that of an Arabian geographer Ibn Khurdabah (869–895 AD), who included Galangal in a list of products from a country he refers to as Sila (which was undoubtedly China). Sometimes later, Plutarch made reference to the use of Galangal as a fumigating spice by the early Egyptians. In 1153, it was mentioned by Edrisi as an incoming spice into Ade, which was then the port used by the Egyptians for all Asiatic products. In his travels, Marco Polo mentions that it was produced fairly extensively in Kachanfu and Kinsai (Tonkin).

Galangal is one of the spices which reached European markets relatively early, for it is mentioned along with black pepper in the literature of the Middle Ages, where it received praise in various writings dealing with drugs, medicines, etc.

The source of the spice which was reaching the European market remained almost unknown until the last century, when in 1867, a plant of unknown botanical origin was discovered in the extreme south of China. The following year, Hance examined both dried and living specimens of the plant and was able to conclude that it was the same as the "Lesser Galangal" known to the pharmacists.

Economic Importance

Kaempferia galanga Linn. popularly known as "*Kacholam*" in Kerala State, India, is a highly priced medicinal and aromatic plant of the family Zingiberaceae. It is a perennial herb, wherein the roots, rhizomes and leaves are aromatic. However, the commercially important part is the rhizome which is used in the ancient Indian medicine, *Ayurveda*, as a stimulant, diuretic and carminative. The active ingredient of the rhizome is an essential oil, which is also used in the manufacture of perfumery.

Table 10.3 Trends in the export of dried Galangal rhizomes from India

Spice	1990–91		1991–92		1992–93	
	Quantity	Value	Quantity	Value	Quantity	Value
Galangal rhizomes and roots	62.3	520.5	72.5	1177.3	73.7	1501.4

Note Quantity in tons; value in INR
Source Spices Board 1994, Spices Statistics 2nd Edition pp. 1–162

Export

According to The Spices Board, Government of India, New Delhi, 1994, Spices Statistics publication depicts the trend of the export of Galangal rhizome/root (dried) during the 3 years ending in 1992–93. During 1992–93, about 74 tons of Galangal rhizome earned INR worth over 1.5million (see Table 10.3).

Area and Production

At present, no regular reliable information/statistics is, yet, available both on the area covered by Galangal and its total production in India, though regular export figures are reported as shown in Table 10.3.

Crop Management

Detailed information on the above topic is lacking. However, the available information is summed up, as under.

Varieties/Races of Galangal

The investigation was carried out at the College of Horticulture, Kerala Agricultural University, Vellanikkara, Trichur District, Kerala State, India, during 1992–93, using five geographical races of Galangal, namely *Koothattukulam, Palghat, Thodupuzha, Ponnukkara* and *Vellanikkara.* The trial was laid out in Randomized Block Design (RBD) with 4 replications.

In an earlier investigation, by Rajagopalan et al. (1989), a Split-Plot design, consisting of a combination of planting time and stage of harvest in the main plot and type of seed material in the subplot with 3 replications, was adopted for the experiment.

Planting

Seed materials were mother rhizomes and finger rhizomes. The mother rhizomes planted during the third week of May and harvested at 6 months of maturity recorded the maximum essential oil and oleoresin yield. The planting time for Galangal is May–June.

Plant Protection

Diseases

Bacterial Wilt of Galangal: *Pseudomonas solanacearum* has recently been reported, for the first time, causing bacterial wilt of *Kaempferia galanga* from Kerala State, India.

Application of Manures and Fertilizers

Farm Yard Manure (FYM) at the rate of 20 tons/ha is applied at the time of planting. Fertilizers are applied at the rate of 50:50:50 (N, P and K, respectively) kg/ha in two split applications, at the time of first and second weeding.

The mean nutrient uptake by the crop was found to be 22.8 kg N, 28 kg P_2O_5 and 36 kg K_2O per hectare.

Harvest

Harvest is undertaken when leaves start drying.

Yield

The geographical races of *Kaempferia galanga* differ significantly from one another in quantum of yield of rhizome and essential oil production. The oil quality remains the same, however, irrespective of the geographical type of the Galangal plant. Selection programs for crop improvement need to be based on rhizome yield and oil content.

Mother rhizomes used as seed material gave higher yield of both essential oil and oleoresin content. Stage of harvest does not influence yield output. The results are summarized in Table 10.4.

The results of the investigation summarized in Table 10.4 show that there is a significant difference between the five geographical races for rhizome yield (fresh and dry), dry weight recovery, oil content and oil yield. Rhizome yield was highest in *Thodupuzha* (V_3) on fresh weight basis. However, with regard to dry rhizome yield, *Thodupuzha* (V_3) was followed by *Palghat* (V_4). This variation is because of the variation in dry weight recovery of the races tested.

The dry weight recovery of the entries differed considerably; the maximum was for *Koothattukulam* (V_1) 32.5%. Variability in the rhizome dry weight percentage exhibited by the tested entries indicated that the water retention capacity of apparently dry rhizomes at the time of harvest varied with the entry tested. Ponnukkara entry (V_4) was found to have the maximum moisture retention power among the five entries, as is evident from its low dry weight recovery of 29.4%. The above facts show that in *Kacholam*, the geographical races differ in field characters like biomass production and water retention power/capacity of the rhizomes concerned at the time of harvest.

Post-harvest Technology

Rhizomes are cleaned after harvest. Thereafter, they are chopped into circular-shaped slices and dried until hard and crisp. Dry weight of rhizomes, drying recovery (%), oil content (%) and oil quality of each type are then estimated.

Table 10.4 Essential oil and oleoresin content as influenced by planting time, stage of harvest and type of seed materials of *Kaempferia galanga* used

Treatments	Essential oil		Oleoresin content	
	Content (%)	Yield (kg/ha)	Content (%)	Yield (kg/ha)
1. Planting time				
Third week May	1.68	19.1	4.9	56.3
First week June	1.74	18.2	4.1	43.1
Second week June	1.69	14.0	3.4	28.5
Critical difference ($P = 0.05$)	N.S	3.7	0.89	9.09
Second-stage harvest				
Sixth month	1.63	16.5	4.3	44.3
Seventh month	1.62	17.9	3.8	43.5
Eighth month	1.86	16.9	4.3	40.7
Critical difference ($P = 0.05$)	N.S	N.S	N.S	N.S
3. Seed material				
Mother rhizome	1.66	18.2	4.2	46.9
Finger rhizome	1.74	17.9	4.1	38.4
Critical difference ($P = 0.05$)	N.S	N.S	N.S	7.5

Note N.S. Not significant
Source Rajagopalan et al. 1989

Composition

The published information on the composition of the rhizome or nutritive value is quite scanty (Rajagopalan et al. 1989).

However, some useful information is available on the volatile content of the rhizome as well as the oleoresin content. These are briefly discussed below.

Volatile Oil

Galangal oil, which is also known as "False Ginger Oil", "Lesser Galangal Oil" and sometimes as "Chinese Galangal Oil", is steam-distilled from the dried comminuted rhizomes of Galangal. Galangal oil is a pale-yellow to olive-brown liquid with an eucalyptus—cardamom-ginger like odor and warming camphoraceous—like bitter taste.

The green rhizomes have the following composition as summarized in Table 10.5.

The essential oil consists of methyl cinnamate (48%), cineol (20–30%), some camphor and probably d-pinene. Leaves also contain a volatile oil (Rajagopalan et al. 1989).

There is a variation in composition of rhizome oil as reported by different scientists, which is naturally due to the variance in the volatile oil itself, the composition of which, in turn, is affected by a number of factors such as the following ones:

1. Area where Galangal is cultivated

Table 10.5 Composition of the green rhizome

Property	Content
Essential oil	0.04%
Specific gravity at 19°	0.978–0.985
Optical rotation	+ 4.6°
nD	1.5164
Ester value	145.6

2. Age of the plant
3. Season and climate conditions
4. Type of soil
5. Time of harvest
6. Method of distillation

Essential oil content is not significantly influenced by any of the treatments of the above-mentioned investigation, such as the planting time and stage of harvest. For essential oil yield per hectare, rhizome planted during the third week of May was superior to the other treatments.

In *Kaempferia galanga*, essential oil is present in the translucent globules (90–110 μm), located in the cortical cells of the rhizomes (Aiyer and Kolammal 1964). The number of oil globules, therefore, determines the essential oil content. Since, it is a genetic character, planting time, stage of harvest or type of propagation material is not expected to have significant impact on the essential oil content.

In essence, it reveals that the geographical races differ only in quantum of oil yield and not on its quality. In other words, the chemical composition of essential oil in *Kaempferia galanga* is independent of the geographical types. This suggests that in screening programs for crop improvement in *Kaempferia*, emphasis should be placed on evaluation of rhizome yield (fresh and dry) and its oil content.

Regarding oil content, TLC analysis of oil revealed the presence of five major components. Furthermore, GLC pattern was similar for all the races tested, indicating that oil quantity is the same.

The above point to the possibility of manipulating these traits and bringing improvement in the selection of geographical races to meet the requirements of the farmers as well as the user industries.

Oleoresin Content

Planting time has a significant influence on the oleoresin content. Crop planted during the third week of May recorded the maximum oleoresin content (4.91%), followed by the crop planted in the first week of June (4.10%). The same trend was noticed with respect to oleoresin yield per hectare also. Among the propagating materials, mother rhizome was superior to the finger rhizome. Time of harvest did not have any significant impact on oleoresin content as well as the total oleoresin yield.

The high oleoresin content in the early planted rhizome might be the result of the favorable photoperiod. Photoperiod has a significant influence on the mobilization of photosynthates and allied products in the rhizome (the sink) (Leopold and Kriedeman 1978).

Uses

Galangal Rhizomes

In the Indian indigenous system of medicine, the rhizomes are used in rheumatism and catarrhal afflictions, especially bronchial catarrh. The drug is a depressant of the cardiovascular system. Respiration in experimental animals is stimulated by small doses, but depressed by large doses. The drug has an important action on the bronchioles. The rhizome and its essential oil are useful in respiratory complications, especially in children. The rhizomes have also carminative and stomachic properties. In Malaysia, they are used as a spice, and the fruits are a substitute for cardamom.

Galangal oil

It is used as a trace constituent in flavor investigations, but is rarely used in perfumery. It could, however, be of value in an oriental or spice-type perfume formulation.

Galangal Oleoresin

It is used in pharmaceuticals/nutraceuticals as a food flavorant.

References

Aiyer KN, Kolammal M (1964) Pharmacognosy of ayurvedic drugs. Kerala Department of Pharmacognosy, University of Krala, Trivandrum, Kerala State, India

Guenther E (1973) The essential oils, 5th reprint. Oil of horse-radish. Rol. E. Kriegar Pub. Co. Hutengten, New York, pp 63–64

Leopold AC, Kriedemann PE (1978) Plant growth and development. Tata Mcgraw Hill Publishing Co. Ltd. New Delhi

Nair KPP (2022) Minor spices and condiments—global economic potential. Springer Nature, Switzerland AG

Sarasan V, Thomas E, Lawrence B, Nair GM (1993) Plant regeneration in *Piper longum* L. (Piperaceae) through direct and indirect adventitious shoot development. J Spices Aromatic Crops 2(1&2):34–40

Stoll and, Seeback (1948) Helv Chim Acta 31:1432

Willstatter and Stoll (1918) Liebigs Ann 416:21

Chapter 11
Quality Control: National and International Quality Standardization for Spices

Abstract The chapter will discuss, at length, the various aspects of quality standardization for spices.

Keywords Quality · Standardization · National · International · Spices

"Quality control" constitutes the very backbone of the success in any industry, in any field of activity, may it be agriculture or spices, since quality is never an accident, but the result of a conscientious effort, as rightly visualized by John Ruskin. The quality of spices is evaluated by the following three types of techniques:

1. Physical or sensory evaluation
2. Proximate physico-chemical composition and determination of chemical constituents
3. Nutritional assessment

   In the following table (Table 11.1), presentation of the five proximate physico-chemical compositions of 5 minerals and 5 vitamins in as many as 35 spices is presented. Since most of the spices are used basically in small quantities in both vegetarian and nonvegetarian food as "Food Additives", they contribute directly very little to human nutrition, but indirectly spices can play a very valuable role in healthy human nutrition by not only helping to increase the appeal and appreciation of the food dishes, but also in various other ways as agents with nutraceutical and pharmaceutical properties. For instance, turmeric has been found to have immense pharmaceutical and nutraceutical properties. Most recently, it has been found to be very beneficial during the COVID-19 pandemic. Excepting in most restrictive diets, spices are regularly used in most food preparations, and they can make a difference which persuades a patient to stay faithful to an otherwise unappealing diet. The analysis confirms that even the highest calorie spice, Poppy seeds, will add no more than 2 or 3 calories per serving in normal usage. This should be good news, in particular, to weight-watchers. Considering the amounts used, the others typically contribute no more than a calorie per serving and usually less than that. Having this definitive data at last, nutritionists and all those interested in good nutrition may utilize spices to their full potential in making wholesome foods as excitingly delicious as they are good for us.

K. P. Nair, *A Compendium of Unique and Rare Spices*,
https://doi.org/10.1007/978-3-031-20249-0_11

Table 11.1 Proximate composition of ground spices (per 100 g edible portion) of those discussed in this book

Spice	A	B	C	D	E	F	G	H	I	J	K	L	M	N	O	P	K
Caraway	6.4	465	21.4	23.1	43.5	10.4	5.6	0.7	50	0.02	1.9	8.5	0.38	0.38	8.1	12	580
Dill seeds	6.6	435	13.1	17.9	56.4	20.7	6.0	1.6	21	0.01	1.1	11.8	0.42	28	2.8	12	175
Poppy seeds	3.7	530	23.1	35.5	30.2	8.7	7.5	1.6	0. 88	0.01	0.8	9.9	0.69	0.18	9 0.9	12	175
Sage	5.7	415	10.2	14.1	62.3	16.0	7.7	1.8	0.09	0.01	1.0	27.3	0.75	0.345.7	39.8	23	95
Savory	9.1	355	7.1	5.2	69.9	15.3	8.7	2.2	0.14	0.02	1.1	37.8	37	0.04	4.1	12	865
Tarragon	4.7	365	24.2	7.3	51.5	6.8	12.3	1.3	0.81	0.07	3.2	35.7	0.25	1.34	8.9	12	175
Thyme	7.1	340	6.8	4.6	68.3	24.3	13.2	2.1	0.20	0.08	0.9	135	0.51	40	4.9	12	175

Source Analyses performed on Research of 900 Laboratories, St. Louis, Mo. USA. ASTA, USA
AGMARK grade specifications for some of the rare and unique spices discussed in this book
Notes A: Water (g), B: Food energy (calorie), C: Protein (g), D: Fat (g), E: Total carbohydrates (g), F: Fiber carbohydrates (g), G: Ash (g), H: Calcium (g), I: Phosphorus (g), J: Sodium (g), K: Potassium (g), L: Iron (mg), M: Thiamine (mg), N: Riboflavin (mg), O: Niacin (mg), P: Ascorbic acid (mg), K: Vitamin A (international units)

Table 11.2 Summary of the data on the list of spices and condiments with AGMARK Grading and Marking Rules as per APGM Act 1937

Spices and condiments	Year of notification for grading	Date of commencement of compulsory grading
Ajowan	1970	01.01.1963
Amchur (sundried and peeled)	1981	–
Raw mango slices/powder	–	–
Asafoetida (compounded)	1984	–
Cardamom*	1962	01.01.1963
Celery*	1967	21.10.1971
Chilies (whole)*	1962	01.01.1963
Chilies (powder)*	1964	01.10.1971
Coriander*	1964	–
Cumin*	1964	03.12.1973
Curry powder *	1956	21.06.1973
Fennel*	1967	21.10.1971
Fenugreek*	1967	21.10.1971
Garlic*	1964	15.09.1965
Ginger*	1964	15.08.1965
Mustard	1964	–
Onions*	1964	15.09.1965
Pepper black*	1969	01.01.1963
Poppy seeds	1964	21.10.1971
Saffron	1973	01.01.1963
Tamarind (seedless)	1972	–
Tamarind (dried)	1983	–
Turmeric[a]	1964	15.08.1965

Source Directorate of Marketing and Inspection, Government of India (1986). Special Compendium
Notes[*] 14 spices of the 22 spices have been covered under the Compulsory Grading Scheme for Export so far under the label AGMARK

Table 11.2 summarizes the data on the list of spices and condiments with AGMARK Grading and Marking Rules framed and notified under the APGM Act 1937

- Data not available

Table 11.3 summarizes data on grade designations and definitions of Saffron quality produced in India

1. Color
2. Final waste content (percentage maximum)
3. Foreign matter (percentage, maximum)

Table 11.3 Summary data on grade designations and definitions of Saffron quality produced in India

S. No.	Grade designation		Special characteristics							General characteristics
1	2	3	4	5	6	7	8	9	10	11
1	Special	Deep red	5	0.5	14	8.0*	1.5*	55.0*	2*	Saffron shall
2	Standard	Light reddish to bright red	15	1.0	14	8.0*	1.5*	55.0*	2*	a: be the dried full cut or broken stigmas of the plant botanically known as *Crocus sativus* Linn b: have the characteristic strong aromatic pleasant and slightly iodinated smell with bitter and slightly pungent taste and be free from any foreign taste or smell especially the musty smell or taste
Nonspecified			1.0%							c: be free from living insects and molds and shall be free from dead insect fragments and rodent contamination, visible to naked eye(corrected to abnormal vision if needed) with such magnification as may be necessary in any particular case and (d): not containing any foreign added coloring matter

Definitions
Floral waste Yellow filaments, pollen, part of ovary and other parts of the flower of *Crocus sativus* Linn. stamen
Notes Foreign matter includes sand, earth, dust, leaf, stem, chaff and other vegetable matter on dry basis, that is, moisture content, and volatile matter free basis as calculated from column 6 matter, volatile at 103 °C. Sulfuric acid diphenylamine test shall be positive, and floral waste content shall be determined by standard method
Nonspecified—Saffron under nonspecified grade shall be exported only against the agreement between the buyers and sellers in a firm order, subject to limitations as mentioned under column 5 above of the schedule
* refers to the high quality of sample

Table 11.4 National PFA quality standards for the rare and unique spices discussed in this book

Name of spice	Botanical name	PFA quality standard
Caraway (whole) *Siah jira*	*Carum carvi* Linn	A.05.01
Poppy seeds (*Khas* whole)	*Papaver somniferum*	A.05.18
Saffron (*Kesari*)	*Crocus sativus* Linn	A.05.19

4. Volatile matter at 103 degree Celsius (percentage by weight, maximum)
5. Total ash percentage (by weight, maximum)
6. Ash insoluble in HCl (percentage by weight, maximum)
7. Aqueous extract (percentage by weight, minimum)
8. Total nitrogen (percentage by weight, minimum)

Table 11.4 summarizes the national PFA quality standards for the rare and unique spices discussed in this book.

International Standards

Development of International Standards (ISO) For Spices and Condiments and their Methods of Test:

The International Organization for Standardization (ISO) is the specialized international agency for development of international standards through its setup of 191 technical committees covering different commodities traded internationally.

The membership of ISO at present comprises 89 countries. The ISO aims to promote development of standardization and related activities in the world with a view to facilitating international exchange of goods and services and developing for cooperation in the sphere of intellectual, scientific, technological and economic activities. The results of ISO technical work are published as international standards (ISO Standard).

One of ISO Technical Committees, ISO/TC 34, deals with agricultural food products. The technical committee has 15 sub committees dealing with various food commodity groups, such as fruits and vegetables, milk and milk products, meat and meat products, spices. One of the subcommittees (ISO/TC 34/SC 7) deals with spices and condiments for which the Government of India has the honor of holding the Secretariat in view of India's importance as one of the major spices producing and exporting countries in the world.

Membership of the Spices Subcommittee (SC 7) has 15 participating members, 24 observer members and 6 international bodies having Category A Liaison and Category Bi Liaison. Under the procedural rules of the ISO, every participating member has an obligation to vote and attend the meetings of the ISO/TC 34/SC 7. The Observer and Liaison members receive all technical documents for comments, but have no voting rights, while they have the right to attend the meetings. A list of ISO Standards published on various spices so far is given in the Table 11.5. The standards are subject to periodical review (once in 5 years) under the working procedure of ISO with a view to keeping the standards up-to-date.

Table 11.6 summarizes the ASTA (USA) cleanliness specifications.

Table. 11.7 summarizes the data on defect action levels for spices, as prescribed by the USFDA.

Interpretation of Insect Filth in Foods

Spice numerical tolerances for insects or insect fragments in foods do not include aphids, mites or thrips unless there are specific references to these insects in the Regulatory Action Guidance section of the Compliance Policy Guide. Tolerances for insects in foods refer to dead insects. Live infestation should be handled in accordance with USFDA Administrative Guidelines 403, 01, Food Storage and Warehousing, USA.

Future Research and Development Needs of Minor Spices and Condiments

India is traditionally known as the "Home of Spices". Until late fifties and early sixties, the country enjoyed a unique position in the world to deserve the title. Even the history of India and its subsequent conquer and rule by the British has the spice trade as a background. Two important spices, black pepper and cardamom, come to the mind. For instance, India virtually controlled the global cardamom trade, controlling as much as 90–95% of the global take off. Not anymore. Things have vastly changed since. Guatemala has emerged as a serious competitor to India, and now India supplies hardly 30–40% of the global cardamom requirement. Similar to cardamom, India used to supply as much as 70% of the global requirement of black pepper. Not anymore. Countries such as Brazil in North American continent and Malaysia and Indonesia in South Asian continent have emerged as serious competitors to India. India's position has tumbled down to more than 25% of the world supply.

Taking into account the above facts, a national symposium was organized in April 1987 in New Delhi, under the guidance of late Dr. J. S. Pruthi, as President of AFST (India) and Chairman of the Organizing Committee of the Association of Food Scientists and Technologists (India), Delhi chapter, in collaboration with the Indian Council of Agricultural Research, New Delhi, India, the Spices Board, National Horticultural Board, Directorate of Marketing and Inspection, Ministry of Agriculture, Government of India, in New Delhi, India, with the following objectives:

1. Reviewing the progress made so far in research and development activities relating to all aspects of spices in the country
2. To identify or determine the research and development needs/gaps still present and finally
3. To make suitable suggestions notably in respect of the following:
    (i) Improvement in agro-horticultural techniques or what is generally known as the "Package of Practices" (POP)
    (ii) Development of new high-yielding, disease and pest-resistant/tolerant and drought-tolerant varieties
    (iii) Modernization of post-harvest technology of spices

Table 11.5 A published international quality standards (ISO and ISO/R) for spices and their methods of test

S. No.	ISO standard	Details
1	ISO 676–1962	Nomenclature of spices and condiments
2	ISO 882–1980	Cardamom specification
3	ISO 927–1982	Determination of extraneous matter
4	ISO 928–1980	Determination of total ash
5	ISO 929–1980	Determination of water-soluble ash
6	ISO 930–1980	Determination of acid-soluble ash
7	ISO 939–1980	Determination of moisture content (Entrainment method)
8	ISO 940–1980	Determination of alcohol-soluble extract
9	ISO 941–1980	Determination of cold water soluble extract
10	ISO 949–1980	Sampling of spices and condiments
11	ISO/R 959–1969	Black and white pepper (whole and ground) specification
12	ISO/R 927–1969	Chilies (whole and ground) specification
13	ISO 973–1980	Pimento (all spice) whole and ground specification
14	ISO 1003–1980	Ginger (whole, in pieces or ground) specification
15	ISO 1108–1980	Determination of nonvolatile ether extract
16	ISO 1208–1982	Determination of filth
17	ISO 1237–1981	Mustard seed specification
18	ISO 2253–1974	Curry powder
19	ISO 2254–1980	Cloves (whole and ground—powdered)
20	ISO 2255–1980	Coriander (whole and ground-powdered)
21	ISO 2825–1981	Preparation of a ground sample of spice for analysis
22	ISO 3493–1976	Vocabulary of vanilla
23	ISO 3513–1977	Determination of Scoville Index for chilies
24	ISO 3588–1977	Determination of degree of fineness of grinding—hand sieving method (reference method)
25	ISO 3632–1980	Saffron
26	ISO 5559–1981	Dehydrated onion
27	ISO 5560–1981	Dehydrated garlic
28	ISO 5561–1981	Caraway (whole)
29	ISO 5564–1982	Piperine determination in black or white pepper (whole or ground) by spectrophotometry

(continued)

Table 11.5 (continued)

S. No.	ISO standard	Details
30	ISO 5565–1982	Vanilla (*Vanilla fragrans*) (Salisbury) Ames
31	ISO 5566–1982	Determination of coloring powder in turmeric by spectrometry
32	ISO 5567–1982	Determination of volatile organic sulfur compounds in dehydrated garlic
33	ISO 6538–1982	Specification of cassia (Type China Type Indonesia and Type Vietnam (whole and ground-powdered) specification
34	ISO 6539–1983	Cinnamon (Type Sri Lanka, Type Seycheles and Type Madagascar) (whole and ground-powdered)
35	ISO 6675–1982	Fenugreek (whole or ground-powdered

B Draft International Standards (DIS) and Draft Proposals for Spices

S. No	International Standards (DIS) and Draft Proposals (DP) for Spices	Details
1	DP 959/1	Revision of ISO/R 959–1969 Part I. Black pepper (whole and ground)
2	DP 959/2	Revision of ISO/R 959/2–1969 Part II. White pepper
3	DIS 972	chilies and capsicum—Revision of ISO/R 972–1969 for chilies and capsicum
4	DIS 5562	Turmeric (whole and ground)
5	DIS 5563	Dehydrated peppermint
6	DIS 6465	Whole cumin
7	DIS 6571	Determination of volatile oil content in spices and condiments
8	DP 6573	Blond caraway
9	DIS 6574	Whole celery seed
10	DIS 6576	Laurel (whole and ground)
11	DIS 6577	Nutmeg and mace
12	DIS 6574	Thyme (whole and ground)
13	DIS 7377	Juniper
14	DIS 7386	Aniseed (whole and ground)
15	DIS 7540	Ground paprika
16	DP 7541	Determination of coloring matter content in ground paprika
17	DIS 7542	Microscopic examination of ground paprika

(continued)

Table 11.5 (continued)

S. No.	ISO standard	Details
18	DP 7543	Determination of capsaicinoids content in chilies
19	DP 7925	Wild marjoram
20	DP 7926	Tarragon
21	DP 7927	Sweet fennel
22	DP 7928	Savory

(iv) Development of new "value-added" spices processed products and "by-products" from spice waste
(v) Development of improved packaging techniques and packages
(vi) Better storage and warehousing techniques
(vii) Improved distribution of marketing of spices and spice products
(viii) Development of new end-uses in culinary flavoring, pharmaceutical, nutraceutical and perfumery and cosmetics industries.

About 80 research and development papers (including 59 research papers and 21 lead or invited papers by the concerned specialists were presented and discussed in this national seminar during two days, based on which it was unanimously decided to request late Dr. Pruthi to compile and scrutinize carefully all the suggestions made during the seminar, edit them and compile them into a compendium, in consultation with the various specialists. The following is a highlight of this effort.

A. Agro-Horticultural Aspects: Production, Productivity, New Varieties, Plant Protection Measures, New Thrust Areas and Priorities
B. Post-Harvest Technology
C. Processing Technologies
D. Packaging of Spices (whole, ground, mixed and other spices-based products)
E. Warehousing and Storage of Spices Spice products and Condiments
F. Quality Control and Standardization (National and International)
G. Shipping and Transportation of Spices
H. Development of Domestic and Export Marketing of Spices
I. New End-Uses and Utilization of Spice Wastes
J. Other Aspects—Diversification and Sensory Evaluation

The following is a brief discussion on each of the items mentioned above.

Agro-Horticultural Aspects

Production, Productivity, New Varieties and New Thrust Areas

Need for yield enhancement of spice varieties: Possibly, the most important aspect of production in India is the enhancement of yield, as, on average, yields in India are the lowest. This raises cost of production at the expense of the farmer.

Vigorous Extension Work on New Agro-Techniques Among Spice Farmers: Though the spice research scientists have been able to evolve better spice farming

Table 11.6 ASTA (USA) latest cleanliness specifications (revised January 1, 1990) for spices

Name of spices, seeds or herbs	Whole insect dead by count	Excreta mammalian (mg/lb)	Excreta others (mg/lb)	Mold by weight (%)	Insect defiled tested (% by weight)	Extraneous foreign matter by weight
All spice	2	5	5	2	1	0.5
Anise	4	3	50	1	1	1
Caraway	4	3	10	1	1	0.5
Cardamom	4	3	1	1	1	0.5
Cassia	2	1	1	5	2.5	0.5
Cinnamon	2	1	2	1	1	0.5
Celery	4	3	3	1	1	0.5
Chilies	4	1	8	3	2.5	0.5
Cloves	4	5	8	1	1	1
Coriander	4	3	10	1	1	0.5
Cumin seed	4	3	5	1	1	0.5
Dill seed	4	3	2	1	1	0.5
Fennel seed	*	*	*	1	1	0.5
Ginger	4	3	3	*	*	1
Laurel leaves	2	1	10	2	2.5	0.5
Mace	4	3	1	2	1	0.5
Marjoram	3	1	10	1	1	1
Nutmeg (broken)	4	5	1	*	*	0.5
Nutmeg (whole)	4	0	0	*	*	0
Oregano	3	1	10	1	1	1
Black pepper	2	1	5	*	*	1
White pepper	2	1	1	*	*	0.5
Poppy seed	2	3	3	1	1	0.5
Rosemary leaves	2	1	4	1	1	0.5
Sage	2	1	4	1	1	0.5
Tarragon	2	1	1	1	1	0.5
Thyme	4	1	5	1	1	0.5

(continued)

Table 11.6 (continued)

Name of spices, seeds or herbs	Whole insect dead by count	Excreta mammalian (mg/lb)	Excreta others (mg/lb)	Mold by weight (%)	Insect defiled tested (% by weight)	Extraneous foreign matter by weight
Turmeric	3	5	5	3	2	0.5

Notes *See ASTA Booklet for details
ASTA American Spice Trade Association
Source 1. M/s. A. A. Sayia & Co. Inc. New Jersey, USA. Weekly Market Report. Vol II No. 53 January 5, 1980

techniques, the weakest link is the extension work in the chain. This focuses on the need to evolve better "farmer friendly" approaches.

Bridging the yield gap: The average national yield for spices is not only much lower than the potential yield, but also the actual realizable yield. Concerted inter-disciplinary programs, followed by vigorous extension work, are immediately required to bridge the "yield gap".

Development of New Varieties

No successful technological innovation or improvement is possible, without the assurance of a steady and regular supply of high-quality raw material. This calls for increased research in the agricultural and horticultural aspects of various spice crops which have not received the attention they deserve. Development of new, high-yielding and disease-resistant varieties and study of their precise degree of nutrition, through intelligent fertilizer management, is an immediate necessity. It is in this context that the relevance of a revolutionary soil and fertilizer management concept developed by this author (Nair 2018), now globally known as "**The Nutrient Buffer Power Concept**", calls for special attention.

Improvement in Active Principle Content and Spice Yield

This aspect of research would be relatively easy with annual spices, but there will be long time lag with spice perennials like cassia, nutmeg, clove, etc.

As the number of low-yielding spice varieties are still large in many countries, efforts should be made to reduce their number being cultivated and attention must shift onto those with high-yielding capabilities.

Planned Production and Marketing

Price fluctuations can only be controlled through effective and planned production strategies

Large-scale cultivation of tree spices, seed spices and leafy spices

Spices like nutmeg, cassia, cinnamon, cloves and leafy spices and herbs like mace, rosemary, sage and seed spices like all spice are in great demand. Planned production must focus on these.

Table 11.7 United States food and drug administration (USFDA) defect action levels for spices

Product/spice	Defect	Action level
All spice (whole)	Mold	Average of 5% or more moldy berries by weight
Bay laurel leaves	Mold	Average of 5% or more moldy leaf pieces by weight
	Insect filth	Average of 5% or more insect-infested leaf pieces by weight
	Mammalian excreta	Average of 1 mg or more mammalian excreta per pound of leaf after processing
Capsicum pods	Insect filth and/or mold	Average of more than 3% by weight of insect- infested and/or moldy pods
Ground capsicum (including paprika)	Mold	Average mold count of more than 20%
	Insect filth	Average of more than 50 insect fragments per 25
	Rodent filth	Average of more than 6 rodent hairs per 25
Ground paprika	Mold	*Average mold count of no more than 20%
	Insect filth	*Average of more than 75 insect fragments per 25
	Rodent filth	*Average of more than 11 rodent hairs per 25
Cassia or cinnamon bark (whole)	Mold	Average of 5% or more moldy pieces by weight
	Insect filth	Average of 5% or more insect-infested pieces by weight
	Mammalian excreta	Average of 1 mg or more of mammalian excreta per pound
Cloves (whole)	Stems	Average of 5% or more stems by weight
Condimental seeds	Mammalian excreta	Average of 3 mg or more of mammalian excreta per pound
Other than fennel seeds and sesame seeds		
Cumin seed	Sand and grit	Average of 9.5% or more ash and/or 1.5% or more acid-insoluble ash
Curry powder	Insect filth	Average of 100 or more insect fragments per 25

(continued)

Table 11.7 (continued)

Product/spice	Defect	Action level
	Rodent filth	Average of 4 or more rodent hairs per 25
Fennel seed	Insects	20% or more of subsamples contain insects
	Mammalian excreta	20% or more subsamples contain mammalian excreta or average of 3 mg or more of mammalian excreta per pound
Ginger (whole)	Mold and/or insect filth	Average of 3% or more pieces by weight are moldy and/or
	Mammalian excreta	Insect-infected or average 3 mg or more of mammalian excreta per pound
Mace (whole)	Mold and/or insect filth	Average of 3% or more pieces by weight are moldy and/or insect-infested
	Mammalian excreta	Average of 3 mg or more of mammalian excreta per pound
	Foreign matter	Average of 1.5% or more foreign matter through a 20-mesh sieve
Nutmeg (whole)	Insect filth and/or mold	Average of 10% or more pieces by count are insect- infested and/or moldy
Pepper (whole)	Insect filth and/or mold	Average of 1% or more pieces by weight are insect-infested and/or moldy
	Mammalian excreta	Average of 1 mg or more of mammalian excreta per pound
	Foreign matter	Average 1% or more picking and sifting by weight
Sesame seeds	Insect filth/decomposition	Average of 5% or more insect-infested or decomposed seeds by weight
	Mammalian excreta	Average of 5 mg or more of mammalian excreta per pound
Spices, leaves, other than bay leaves	Foreign matter	Average of 0.5% or more foreign matter by weight
	Insect filth and/or mold	Average of 5% or more pieces by weight are insect-infested and/or are moldy
	Mammalian excreta	Average of 1 mg or more of mammalian excreta per pound, after processing

An important suggestion: There is an urgent need to survey, collect, conserve, evaluate, maintain and distribute the germplasm collections of all important spices in a systematic manner.

Need for evolving varieties with specific quality attributes

The emphasis in breeding spice crops must shift exclusively from yield to quality standards, keeping in mind the market needs, both native and international.

Need for more research and development work on seed spices, such as ajowan, aniseed, caraway, celery, coriander, cumin, dill, fennel and fenugreek.

Need for introducing into India high-yielding accessions from the Mediterranean regions and their related R&D work.

The following are the production constraints in the above group of spices:

1. Slow germination
2. Poor establishment
3. Lack of adequate and consistent response to application of fertilizers and manures
4. High incidence of diseases like "wilt", "blight" and "powdery mildews"
5. Lack of frost-resistant varieties in coriander and fenugreek

Need for Special Attention on R&D work on Tree Spices

Tree spices, such as clove, nutmeg and cinnamon, need to be given special attention with respect to area, production, productivity, post-harvest technology, quality and processing, since India is a net importer of these spices, leading to much drainage of hard currency. This can be plugged if the country focuses more on these tree spices.

Other Spices

Saffron is the costliest spice in the world. Kashmir Saffron is as good as the Spanish variety. However, the Kashmir Saffron yield is much lower compared to the Spanish Saffron. In Spain, there is now considerable reduction in acreage of Saffron, as it is a labor-intensive crop and migration of labor from rural areas to cities has adversely affected its cultivation in Spain. India will do well to capture this possibility to focus on Kashmir Saffron and consequently capture world market, which will be a boon to Indian economy. Saffron is discussed in great detail in this book.

Developmental and Extension Services on Agro-Horticultural Aspects.

There is an urgent need to provide the infrastructural facilities for the commercial-scale production of the spices planting material of the recommended new spice varieties.

Plant Protection Measures

Suggestions for plant protection are covered under the following three subheads:

1. Diseases
2. Insect pests
3. Nematodes

Diseases

Spices are affected by any diseases which are prevalent in Kerala, Tamil Nadu, Karnataka, Rajasthan, Gujarat, Andhra Pradesh, Himachal Pradesh, Sikkim, Bihar and Uttar Pradesh. The relative seriousness of the individual disease varies from state to state. There is, therefore, an urgent need to intensify research in the following areas:

(i) Scientific estimation of the crop losses due to at least the major diseases
(ii) Thorough understanding of the epidemiology
(iii) Collection and utilization of meteorological data
(iv) Evaluation of cultivated germplasms for resistance against diseases
(v) Introduction of exotic collections, especially in cumin and coriander
(vi) Phytosanitary restrictions in seed movement
(vii) Proper maintenance of disease-free nurseries
(viii) Popularization of pre-sowing and pre-storage phytosanitary measures
(x) Minimize costly plant protection measures
(xi) Adhere to Central Insecticides Board (CIB) recommendations
(xii) Need to initiate studies on pesticide residues in soil
(xiii) Need to test efficacy of testing eco-friendly practices like crop rotation in disease control
(xiv) Need for laying out large demonstration plots for farmer benefit, by utilizing facilities in Krishi Vigyan Kendra
(xv) Effective investigation on efficient storage facilities to minimize post-harvest losses in storage in crops such as onion, garlic, leafy spices, large cardamom, cumin and fennel

Insect Pests

Following are some of the areas requiring urgent attention:

(i) Identification of the most important field and storage pests
(ii) Quantify injury level to the plant/product
(iii) Proper survey of insect pests like shoot borer, thrips and scales
(iv) Bionomics and control of scales and other pests
(v) Survey of storage pests and investigation of residue analysis in all seed, leaf, bulb and tree spices

Nematodes

The following are the critical areas demanding attention:

(i) Investigations, in particular, in nutmeg, cinnamon and clove on nematode infestation
(ii) An integrated approach for utilizing (a) resistant/tolerant and productive genotypes, (b) cultural practices involving organic amendments and (c) with minimal use of agro-chemicals which can keep the nematode population under economic threshold levels, which will ultimately lead to greater productivity as well as total production.

(iii) Intensifying germplasm screening to identify genotypes with horizontal resistance to major nematode infestation in various spice crops
(iv) The ideal oilcake/nematode combinations which keep the nematode population under check. Investigations on Vesicular Arbuscular Mycorrhizae (VAM) in nematode control
(v) Identification of biological control measures.

Mycotoxins in Spices during storage

Occurrence of aflatoxins B_1, B_2 and G_1 in 5 out of 10 coriander samples and also during storage has been noted. These investigations must be done.

Processing Technologies

The following aspects must be investigated:

1. Spice grinding
2. Modern pre-crushing and milling
3. Designing and fabrication of machinery for cryogenic or chilled-water milling of spices
4. Popularization of indigenous cold water milling machinery
5. Packaging and storage investigations on cryogenic milled spices
6. Feasibility studies on anticaking agents and antioxidants during milling
7. Blending of oleoresins from spices, essential oils or synthetic oleoresins or both to develop new flavors to suit market needs
8. Promotion of industrial use of spice oleoresins.

Supercritical CO_2 Gas or Fluid Extraction of Flavor/Oleoresins (SCGE)
Encapsulation of Oleoresins.
Newer Spice Products: The following areas must be investigated.

(i) Spice products, oleoresins and essential oils have great internal and international demand. Systematic investigations on thesemust be initiated
(ii) A continuous surveillance of changes in technology, production and quality
(iii) The different processing techniques adopted by farmers
(iv) Product utilization
(v) Proper post-harvest storage

Packaging of Spices (whole, ground and blends)

Following areas must be investigated:

(i) Flexible and semi-rigid containers
(ii) Use and desirability of plastics
(iii) Use of coextruded thermoplastic multilayer films and bottles.

Packaging of Specialty Products

As a means of enhancing shelf-life of foods, in general, andspices and condiments, in particular, packaging techniques andmaterials have gained much importance. The development ofnew and improved transparent plastic films, foils, laminations,high-speed film-sealing machines, decorators, gas flushers, etc.has opened up new avenues for research and development. This area must be vigorously pursued further.

Unit Packages, Bulk Packages

In these investigations, an important factor to be considered is that in choosing packing materials and the concerned spiceproducts, the migration of the components from the packing materials to the product and *vice versa* must be taken into full consideration.

Other Packaging Requirements

The following must be investigated:

(i) High-density polythene films
(ii) Co-extruded films
(iii) Use of thin-walled pouches/containers
(iv) Efficacy of multilayer films of HDPE/LDPE/Nylon in combination with ionomer resins, which also permitpackaging of edible oils, ghee (Indian clarified butter) oils, must be thoroughly investigated
(v) Efficacy HDPE/PP woven bags, which are one of the proven media for packaging, must be further investigated.

Rigid Containers

Glass containers: The glass containers are becoming popular in spice storage. However, the levy on glass containers is disproportionate (36.75%), and this must be further investigated and examined to rectify.

Tin Cans

R&D efforts should be stepped up to produce tinplate of desired quality for OTS cans, so as to stop the import of tin plates into India and thus avoid the drainage of hard currency.

Packaging Systems

The Government of India must focus on the following three items:

(i) The plant and machinery for the manufacture of HDPE/PP woven bags is also made now locally. The Government of India should support indigenous industry by giving full backing to develop advanced technology and restrict import of machinery
(ii) Today the basic plants to manufacture woven bags, namely tape plant with "T" DIE System (Cost Film Process) and circular looms and lamination plant including Tandem-coating plants, are available indigenously. The Indian Government must fully encourage this enterprise

(iii) The Extrusion Lamination Plants, which are manufactured indigenously, are capable of laminating on different substrates like paper board, aluminum foil, polyester film, etc. These are being used by some food processing industries, including spice processing countries. The Government of India must provide full economic encouragement to this enterprise.

Quality Control and Other Aspects of Packaging:

(i) A high power body of experts should be formed to recommend different packing materials for various types of processed spices (whole, ground and mixed)
(ii) Based on research reports, data on suitability of different packaging materials for various types of processed foods should be compiled for the benefit of the industry. Though ISI has already given a lead in bringing out an Indian Standard (IS: 10171–1982) as a guide on suitability of plastics for food packaging, detailed information must be collated for their use for spice products
(iii) Similar guidelines, as above, are also required for use of co-extruded composite films for packing spice products
(iv) There is considerable scope for R&D work on the determination of shelf-life of various processed spice packets in suitable co-extruded composite films (spices whole and ground, mixed *masalas,* etc.)
(v) Sample packaging systems are required for catering to rural-based technology for various spices and its forms, such as whole, ground, mixed
(vi) Regional testing centers/field testing centers for packaging materials should be strengthened.

Standardization of Packaging Materials: There is also a need for the following:

a. Standardization and grading of packing materials
b. Reduction in the variety of containers
c. Nationalized net weights of packages and their tolerance
d. Objective methods of assessment of packages testing.

Warehousing or Storage of Spice and Condiments

The following points must be carefully considered:

(i) Mold is the most serious problem insofar as stored spices are considered. There is thus an urgent need to properly educate spice producers to the great importance of proper drying before packing
(ii) There is an urgent need to identify potential centers where spices in large quantities are available for storage and to design and construct specialized warehousing facilities to suit specific requirements of different spices. Various agencies like the Spices Board, Central Warehousing Corporation, etc., can help in the above task
(iii) There is a dearth of enough published scientific information on the physical and chemical changes in spices during commercial warehousing during different seasons. This issue must be addressed
(iv) Systematic investigations should also be conducted on the shelf-life of spices.

Microbiological Aspects

Although some useful information on the nature and extent of microbial contamination and changes in microbial load during milling and processing operations is available, there is an urgent need for the development of efficient and economically feasible techniques for the inactivation of microflora in spices without significantly adversely affecting the quality of the product, in particular, aroma and flavor. This issue must be urgently addressed.

Insect Infestation and Its Control

Pilot studies on the effect of treatment of burlap bags to provide protection from insects and microbial infestation during storage and shipment should be undertaken on a large scale of shippers at the international level, since the treatment could be of immense value to the spice industry.

Fumigation of Spices

Atmospheric or vacuum fumigation and serial fumigation of spices on a large-scale should also be undertaken.

Simple methods of Decontamination

Efforts should also be made to develop simple, efficient and reliable methods of decontamination of spices and packed spices, as well as methods of determination of pesticide residues.

Quality Control and Standardization at the National and International levels

National and International Quality Standards

The following measures must be urgently attended to:

(i) In view of the expanding global spice market, along with the stringent quality control measures, spice producers and exporters must be properly educated so that no lapses occur due to which an export market destination could be lost to the nation
(ii) There is an urgent need to introduce microbiological standards in grade specification of spices
(iii) Potential revision of AGMARK specifications in consultation with PFA, ISI, BIS, trade, producers and convenors, once every 5 years
(iv) Separate AGMARK/PFA standards should be laid down for different types of ground spice mixtures, such as *masalas,* (Indian curry mixture powders) for instance, *sambar, rasam* (very popular Sooth Indian vegetarian recipes) and chicken *masalas*
(v) The possibility of introduction of the Compulsory AGMARK Quality Control of selected essential spices, such as turmeric, chilies, coriander and black pepper in a few cosmopolitan cities like Mumbai, Kolkata, Delhi and Chennai
(vi) There is an increasing demand in importing countries for greater quality assurance of spices, especially in respect of hygiene and microbiological aspects

in relation to spoilage and health hazards. Greater attention should be paid to these aspects by the spice export houses and the AGMARK authorities (DMI)

(vii) Contamination of spices by insects, insect fragments and pesticide residues is being viewed with increasing seriousness. Irradiation can provide the answer to all these problems, at least for export purposes, which has recently been permitted by the Government of India

(viii) Difficulties are experienced in getting precise information on the physical, chemical and microbiological standards applicable to import of spices and spice products in importing countries. These may be compiled with the help of ITC (International Trade Center, Geneva, Switzerland)

(x) In order to establish uniform international standards, wherever necessary, to upgrade the existing specifications, considerable first-hand, physical, chemical and microbiological data should be collected for spices in different seasons of the year to help framing more realistic standards

(xi) Thirty-seven countries, including India, have cleared for safety standards for the use of irradiated spice. Both consumers and stakeholders must be made fully aware of this development.

Use of Modern Analytical Techniques and Instrumentation

The following aspects must be taken into consideration:

(i). There is a great need for the application of specific, simple, analytical techniques such as TLC and colorimetry including the determination of the characteristic florescence of spices

(ii). Simple, rapid, reliable and inexpensive analytical techniques should be developed for the determination of the active principles in spices, such as piperine in black and white peppers, capsicum in red pepper, pungent principles in celery, garlic, ginger and onions, spice oleoresins, coloring powder and pigments in red pepper, saffron and turmeric, phenols in nutmeg and mace and sulfur compounds in different spices, such as asafoetida, mustard, garlic and onion

(iii). Similar techniques should be developed for detection of different adulterants in ground spices and spice mixtures, liquid and instant spices.

Spice Flavorant Evaluation

This area must be developed since our knowledge of spices in relation to their flavor quality is rather fragmentary.

Need for International Collaborative Analytical Research Program and Liaison

The FAO/ITC can take initiative in promoting active collaboration between several spice stakeholders, such as scientists, exporters and importers, advisory agencies.

Arbitration in Quality Disputes

Suitable methodology for settling quality disputes between exporters and importers through arbitration should be evolved. The Spices Board may work out the modalities in this regard in consultation with the ITC and other concerned bodies.

Shipping and Transportation of Spices

With regard to the above topic, the following aspects should be taken into consideration:

(i). Spice product transportation is very important because during transportation, much damage can be done to the produce leading to loss of quality. The Spices Board must provide guidelines to growers, traders and transporters, whether national or international
(ii). There is insufficient information on quality deterioration during transport. Relevant R&D efforts must be initiated in this aspect
(iii). Better hygiene must be provided in ship-holds to pre-empt future quality loss due to contamination
(iv). Inadequate sailings, refusal of space and shunting out of cargo on alleged grounds of incompatibility of spices and spice products with other cargo or lack of proper storage have in an increasing measure, affected exports, which should be immediately tackled by the Government of India.

Marketing: Development of domestic and international markets

The following are the most important suggestions:

A. Need to organize Spice Growers Cooperatives: Marketable surplus of spices can vary from 82.8% for ginger to 99% for cardamom. Hence, the problems of marketing of spice products are completely different from that of annual crops like wheat or rice which are bought by the government at the stipulated Minimum Support Price. This leads the spice growers to market their produce, immediately after harvest to the itinerant merchants or in the village. In short, this leads to exploitation of the spice farmer. Because of long gestation period, farmers find it difficult to invest money for the development of spice plantations. Hence, there is an urgent need to organize spice farmers co-operatives to integrate production, processing and marketing activities to reap the benefits of spice growing, with appropriate economy of scale.
B. Need to organize and develop transport and communication facilities: Many of the spice-growing areas are located in the very interior pockets of India where market accessibility is very poor. There is a need to organize and develop transport and communication facilities in such remote areas.
C. Need to regulate production and supply of the products through Spices Board: Demand for spices is inelastic. Over production creates glut in the market leading to price nosedive. Even for spices like black pepper and small cardamom, which are almost entirely exported, overproduction has serious economic consequences, because of international market price fluctuations. Establishment of the Spices Board helps the spice farmers to a large extent in this context.
D. Need for Institutional Marketing Arrangement to protect the interests of spice growers and consumers: Marketing of spices is dominated by the private trade with a large number of intermediaries. Producers get only a small share of the price paid by the consumer. Apart from the role played by the National Agricultural Cooperative Marketing Federation of India (NAFED), in procurement and

price stabilizing operations in ginger, and Spices Board for cardamom, there are no institutional arrangements for protecting the interests of the producers-sellers and consumers of spices. The role of institutional marketing in spices should be extended as spices are a very valuable foreign exchange earners.

E. Need for Market Intelligence and Market Regulation of All Spices: The marketing of spices in the country suffers from a lack of market intelligence, absence of market regulations, unfair trade practices, business malpractices, delayed payments to producers-sellers, and finally this unhealthy scenario adversely affects the consumer. As per the recommendations of the National Commission on Agriculture, all spices should be notified for regulation of marketing and the necessary coverage should be provided in the marketing information system.
F. Extension of Voluntary and Compulsory Grading to other important spices: Virtually, very little or no grading is being done in spices at the producers' level. However, selected spices are covered under the Compulsory Export Quality Control and Pre-shipment Inspection Scheme of the Directorate of Marketing and Inspection, for spices, such as black pepper, cardamom, chilies, ginger, turmeric, coriander, curry powder, cumin seeds. The program of voluntary grading as well as compulsory grading for exports should be extended to other important minor spices and condiments, as well (Nair 2022).
G. Stabilization of Production and Ensuring Competitive Prices in Export Markets: The processing facilities, market intelligence and research should be undertaken to stabilize production and to ensure competitive prices to the producer in the export markets.
H. Role of Central and State Cooperative Organizations: When supply exceeds demand, the state and central agencies-controlled cooperatives should enter the market and procure the surplus. This could be maintained as a buffer stock to be released in the market during the off-season.
I. Introduction and Enforcement of APGM Act: The Agriculture Produce and Marketing Act (APGM Act) should be introduced and enforced in states such as Kerala and Karnataka, where large black pepper and cardamom plantations are situated. In areas, where this is already existing, necessary facilities for drying, grading and storage should be built up. The farmers should be educated for selling their produce in regulated markets.

   With the advent of the internet, e-markets are now emerging which enables the farmer to reap the benefit of the best prices for his produce.
J. Need for Soliciting Co-operation of Public Institutions: Steps must be taken to obtain the co-operation and participation of public institutions, such as the Food Corporation of India, State Civil Supplies Cooperatives, in order to improve the marketing of these spice crops.
K. Need for strengthening existing Co-operative Marketing Systems: It is necessary to strengthen existing cooperative marketing societies and establish new ones, where they do not exist, to provide adequate and timely credit to producers to free them from the strangle hold the shylocks (money lenders) and commission agents.

L. Boosting exports by diversification of value-added spice products: Boosting export trade by diversification of value-added products like essential oil, oleoresins, powders, curry powders and other spice mixtures.
M. Need for updating marketing reports on spices (whole, ground and mixtures). The Directorate of Marketing and Inspection may update all their published marketing reports on spices by conducting fresh surveys. The reports have been quite useful.
N. Need for regular compilation of spice statistics. Statistical data may be regularly compiled on area, production and inter-state movement of at least 25 important spices. So far, only 8 spices are being covered.
O. Study of quality changes in spices during shipping and storage. There is an urgent need for studies on the changes in spices during shipping or transport to distant destinations in importing countries and also during storage as well, for development of their remedial measures and other post-harvest techniques.

Arresting Fluctuations in Prices

The following are the important considerations:

(i) Fluctuations in prices and production of spices are constraints affecting exports. To ameliorate the situation, the following suggestions are made:

    a. To implement planned production by a centralized agency
    b. To announce a minimum support price for all spices, as in the case of cereal grains, such as wheat and rice
    c. To process surplus stock of spices into value-added products
    d. To make aggressive efforts to market them within the country and overseas

(ii) When supply exceeds demand, the state and central-level cooperatives should enter the market and procure the supplies to stabilize prices. This could be used as a buffer stock to be released in the off-season
(iii) Steps may be taken to obtain the cooperation and participation of public institutions like the Food Corporation of India, State Civil Supplies Corporations in order to improve marketing
(iv) Strengthen existing marketing societies and establish new ones, where they do not exist to provide adequate and timely credit to producers to free them from the vice grip of the money lenders and commission agents
(v) Diversify value-added products to boost export and improve production techniques to launch value-added products, such as the ones developed by the Central Food Technology Research Institute (CFTRI), Mysore, India, for processing turmeric
(vi) The Directorate of Marketing and Inspection may update all their published marketing reports on spice by conducting fresh surveys and publishing timely concise reports
(vii) Improved methods of post-harvest technology of spices developed at different institutions/establishments should be adopted to prevent mold formation and to ensure quality products therefrom.

Stepping Up International Marketing of Spices

Marketing of spices and spice products depends largely on the rapport established between the producer and the marketing agencies. It is highly necessary that production be tailored to suit the requirements of the markets. Suitable steps may be taken to assess the market requirements of the spices at the national and international level so that production of spices can be adjusted accordingly.

Global Market Research

Global demands for the different spices should be projected in a national and dependable context over a length of time, based on systematic market research. Market intelligence, which covers collection and compilation of statistical data on various aspects of spices, product development, market development and sales promotion, needs to be streamlined and improved. Sales and study teams should regularly visit the existing potential world markets in order to help regulate the factors of supply and demand.

New End-Uses and Utilization of Spice Wastes

Spices as Food Additives

Recently, the western world has shown an increasing interest in new flavoring agents. With the growing anxiety about the toxicity of synthetic food additives, emphasis is shifting toward the discovery of newer flavorants, colorants and texturing agents of natural origin. There is a vast scope for exploitation in this new wave of interest. Several herbs and plant materials have been used for such varied applications in different parts of the world, but have not yet received serious attention. Collaborative research programs should be initiated to survey such plant materials for essential oils, oleoresins, pigments, gums, mucilages or polysaccharides. In fact, portions of spice plants themselves should be investigated for such applications. Residues left after extraction of the active principles, in some cases, like chilies, have already been proposed for alternative uses.

Biochemical and In-depth investigations on Physiological Aspects

Although scientific literature on various biochemical and physiologic processes of some major spices of international importance is voluminous, further research is necessary on some of the important areas mentioned below.

Antioxygenic Properties: The exact mechanisms of the antioxygenic properties of spices, their precise mode of action and the actual nature of their contribution in exerting a preservative action in some foods need further elucidation.

Antibacterial Activity of Spices

Investigations should be initiated into the precise mode of action of volatile constituents and spice essential oils *vis-à-vis* the relation between their chemical structures and antibacterial activity.

Medicinal and Physiological Aspects

Most of the spices that are being used in human dietary possess different physiological and medicinal properties. Very little information is only available on the exact mechanism of their physiological action. Some of them are antibacterial. This obviously suggests that they may exert their beneficial effect by suppressing the growth of the harmful intestinal bacteria, but considering the small quantity of the spices consumed, the medicinal effects they exert on the human system seem to be out of proportion to their antibacterial properties. It appears, therefore, that there may be an alternative mechanism of physiological action by which the spices act as therapeutic agents. Further, in-depth investigations are necessary to explore this important aspect. The German and Indian groups of researchers have made a good start in carrying out a series of research investigations on the physiological and medicinal aspects of some important spices like garlic, ginger, pepper, chilies, cloves, etc., commonly used in flavoring of the foods, but there is still considerable scope for intensification or researches in several areas, such as the following:

(i) Prior to 1955, physicians had warned patients with peptic ulcers to avoid spices and highly seasoned foods, without presenting any scientific evidence to support their recommendation. This approach needs thorough investigation. Some attention has been paid only to paprika, pepper and mustard. Other important spices require similar investigation. The proposed concept of a defensive mechanism—the gastric-mucosal barrier—also should be investigated further

(ii) The "circular activity" of the other spices not investigated, so far, may have to be examined. It will be necessary to determine whether "spice effects" depend on the composition of basic food. In any case, it is possible, in principle to effect, by the use of spices, the required changes which cannot be produced by any of the known medicaments

(iii) Extracts of the preparations from spices such as coriander have been reported to improve the utilization of glucose, as shown by the glucose-tolerance tests. These preparations should be tested on human subjects with diabetes (*Diabetes mellitus*). Likewise, it has also been reported that intake of the powder of dried leaves of *Tejpat* (*Cinnamomum tamala*) in small quantities lowers blood glucose level significantly in diabetic patients. Similar reports on the combined beneficial aspects of turmeric powder, fenugreek and *amla (Phyllanthus emblica)* on diabetic patients need more scientific scrutiny. Further, the beneficial effect of turmeric extracts in gastric disorders, application of garlic, celery seed and ajowan seed oil in cases of rheumatism and use of numerous spices believed to possess emmenagogue and/or abortifacient properties need to be verified scientifically

(iv) The diuretic effects of extracts of caraway, anise and sweet fennel seeds have been reported. Similar research should be done on other spices

(v) Further investigations should be initiated to elucidate the nature of the "active principle" in parsley that has been reported to have a tonic effect on uterine musculature

(vi) Garlic causes clinical improvement in the treatment of patients with leprosy. However, its mode of action is not known yet. It is not clear whether the curative property is due to its antibiotic principle, allicin, or to the allithiamine formed as a result of allicin having combined with thiamine or to the "thiamine status" of the patient. These aspects merit further investigation

(vii) Further investigations should also be undertaken on the effect of spices on the adrenocortical functioning

(viii) Onion is beneficial to everyone, irrespective of age, and an effective agent for preventing atherosclerosis in patients with high blood pressure and with violations of secretory functions of the intestines. Similar investigations are needed on other *Allium* species and other spices

(ix) The precise mode of action of spice essential oils and their aromatic chemicals in perfumes and as fungicides and pesticides merit further investigations

(x) Similarly, the precise role of bacterial and enzyme systems in affording protection to the human skin by spice essential oils and aromatics against microbial infections should also be thoroughly investigated.

Search for New Applications and to discover New Uses, New Strategies to Boost Demand for Spices globally

It is the responsibility of the spice producers and the spice traders to discover new strategies to boost the demand for spices. In the field of spices and flavoring extracts, such a need is already apparent. It may be worthwhile to explore the possibility of formulating new beverages (alcoholic and nonalcoholic), snacks, stuffing and a variety of such products. Investigations on the pharmacological properties of the active principles in spices may suggest new applications, as well.

To promote such newer food applications, the following may be possible:

(i) To improve flavor characteristics
(ii) To introduce modifications in spice concentrates
(iii) To suppress discordant notes
(iv) To impart dispersibility.

Basic investigations on the chemistry and chemical alterations may help in achieving these objectives.

International Collaborative Research and Liaison

Research activities in the field of spices can be fruitful only if a close Liaison is established among growers, processors, exporters and export promotion authorities on the one hand and scientific research organizations and universities on the other. Such a well-knit inter-relationship and a continuous feedback of information from exporting and marketing organizations to scientists are absolutely essential. The FAO or ITC could take the initiative in setting up a committee to promote international marketing and research programs on sound scientific lines for the benefit of the spice consumers, globally, and for the betterment of humanity, at large.

Other Aspects

The following also merit consideration:

(i) Further research on spices is absolutely necessary to provide new opportunities in the diversification of trade in spices, spice blends and their products and to find new uses of spices in human physiology
(ii) Our knowledge of spices in relation to evaluation of their flavor quality is fragmentary, to say the least. Flavor is a conglomeration of many contributing factors. We know more about these factors than what we did earlier. However, further knowledge must be continuously acquired as a first priority in the various disciplines engaged in this complex field of spice flavor, if we are to develop methods for its correct and rapid evaluation
(iii) There have been very rapid developments in the area of instrumentation. From the early colorimetry to atomic absorption spectrometry and later electro-ultrafiltration (Nair 1996), newer techniques have come to the fore. Paradoxically, such developments have been confined only to specific fields like soil and/or plant analysis. But, a hand-holding must happen where experts in these fields must come forward to join the spice researchers and advance further knowledge in this vast "spice science".

References

Nair KPP (1996) The buffering of plant nutrients and effects on availability. Adv Agron 57:237–287
Nair KPP (2018) Intelligent soil management for sustainable agriculture. The nutrient buffer power concept. Springer Nature, Switzerland
Nair KPP (2022) Minor spices and condiments—global economic potential. Springer Nature, Switzerland